TECHNOLOGY IN AMERICA

TECHNOLOGY IN AMERICA

A History of Individuals and Ideas

Second Edition

edited by Carroll W. Pursell, Jr.

The MIT Press
Cambridge, Massachusetts, and London, England

For my brother
Tim Pursell

Third printing, 1994

Copyright © 1981, 1990 by The Massachusetts Institute of Technology

This book was printed and bound in the United States of America.

Library of Congress Cataloging in Publication Data

Main entry under title:

Technology in America.
  Bibliography: p.
  Includes index.
  1. Technology—United States—History—Addresses,
essays, lectures. I. Pursell, Carroll W.
  T21.T43   609.73   81-8364
  ISBN 0-262-66049-0 (paper)   AACR2

CONTENTS

PREFACE TO THE SECOND EDITION

During the eight years since this book was first published, it has found a useful niche in the college and university classroom. The experience of almost a decade has shown that while the book does not do everything (there is still room for a number of interpretive textbooks and certainly a collection or two of primary documents, for example), it is a tribute to the quality of the authors' scholarship that the book continues to give a fair representation of the way in which the history of American technology is viewed.

I have taken this opportunity to add three more chapters covering the topics of scientific management, electronic communications (through the example of the loudspeaker), and the cluster of electronic devices as well as institutional arrangements that came together to create the computer industry of the Silicon Valley. It is suggestive that the book's first essay, by Brooke Hindle, and its last, by James C. Williams, highlight the social contexts in which a large number of individuals worked to produce and improve the nation's technology. Perhaps the Golden Age of the American inventor lies somewhere between.

The list of books for further reading has been thoroughly revised to include titles that have appeared since 1981. There are, of course, many more that could have been added, but the interested reader can find a start here for further study.

Carroll Pursell
Case Western Reserve University
Cleveland, Ohio
June 8, 1989

PREFACE

The score of essays in this volume focus on the spread and elaboration of American technology, and on the men and women who shaped this great human activity in its American context. The biographical organization is not intended to convey the argument that history moves forward only through the deeds of Great Men and Women. At the same time, it is true that technology is a fundamental form of human behavior and that ultimately it is people who have the ideas, design the machines and processes, build the institutions, and suffer the costs and benefits of technological change. The people here focused upon include some of the best known in the pantheon of American folk heroes, but their lives serve primarily as windows through which we may observe the interplay of ideas and institutions that combined to shape the tools of their times.

The essays divide roughly between the nineteenth and twentieth centuries. After a brief introductory essay, a chapter (properly highlighting an occupational type rather than a single individual) describes the technology of America's Wooden Age. This is followed by a discussion of Jefferson's explicit perception of the role of technology in a democratic society, and then the introduction of the Industrial Revolution into the young nation and the nation's fundamental contribution to that great epoch: the American System of Manufactures of Eli Whitney and others. The work of Thomas P. Jones illustrates the institutionalization of this industrialization in ongoing educational forms. Cyrus Hall McCormick epitomizes the spread of the industrialization to agriculture, the Garden of Jefferson's yeomen farmers, and James B. Eads allows us to explain the way in which capital and engineering combined to build a transcontinental business civilization of cities linked by transportation networks.

The career of James B. Francis introduces the beginnings of a scientific technology that has, in our own times, come more and more to crowd out traditional, empirical ways of doing things. Bell's telephone was only one startling application of electricity, a science made into an industry by the hands of such as he and Thomas Alva Edison, perhaps the most important transitional figure in the history of American technology. As the story of George Eastman makes clear, the use of science to transform technologies, which can then be made and marketed by large corporations, became a salient feature of twentieth-century American culture.

Ellen Swallow Richards stands as a reminder that technology touches women in our society in many ways differently than men and that the domestic workplace is just that—a place where people (mostly women) work, with or without machines and industrial standards of efficiency. Just as Richards stood for the science of a controlled (domestic) environment, Gifford Pinchot pioneered this Progressive reform among the natural resources of the nation, giving birth to the first Conservation Crusade and the ideal of efficient utilization. The use of science to rationalize various environments was nowhere more dramatically displayed than in the automobile factories of Henry Ford. Harnessing Scientific Management to the American System of Manufactures, Ford's assembly line became the hallmark of the modern industrial age just as his Model T automobile became the representative artifact of modern American civilization.

The tensions inherent in trying to revolutionize American life in the name of traditional values, to guarantee political democracy with an increasingly authoritarian technology, were dramatically symbolized by the first transatlantic airplane flight accomplished by Charles Lindbergh. It was not lost on everyone that the Lone Eagle was entirely dependent upon the very rational and structured industrial order which his flight was thought to transcend. In their own ways Buster Keaton and Charlie Chaplin, classic heroes of Hollywood's Golden Age, also helped the nation confront what it meant to live in the modern technological age.

The development of cheap, flexible, divisible energy in the form of electricity was a consistent thread running through the career of Morris L. Cooke: a goal that promised to bring the blessings of modern technology within the reach of all Americans. But already Enrico Fermi, through his work on nuclear power, and Robert H. Goddard, with his pioneer experiments in rocket flight, were pointing the way to new technologies so steeped in the mysteries of science, and of such a scale as to outstrip the budgets of many nation states, that both could only enhance corporate (both public and private) as against individual power.

These essays first appeared in a volume published abroad by the Voice of America and were based upon a series of overseas broadcasts designed to celebrate the nation's bicentennial. Richard J. Gordon, at that time Forum Editor of the Voice of America, originated the idea. Carol Morarsky and Merrill Miller were also extremely helpful in organizing the project and bringing it to fruition.

Carroll W. Pursell, Jr.
University of California, Santa Barbara
February 1981

1

TECHNOLOGY IN AMERICA: AN INTRODUCTION

Carroll W. Pursell, Jr.

The year 1776 saw two great revolutions unfolding. In Europe the Industrial Revolution was beginning to change both the face of the earth and the lives of people everywhere. In America a political revolution was bringing forth a new continental empire. The revolutionary generation of Americans was born in a country which labored with a medieval technology in a colonial economy. By the time its work was done, it had deliberately borrowed the new European industrial technology and rejected the Old World's political control. Inevitably those two revolutions became intertwined and both were changed. American innovations produced mass production and enabled mass consumption. Industrial technology directed American energies toward economic growth and continental expansion. Only half consciously, the United States followed the advice of the "Father of Scientific Management," Frederick Winslow Taylor, to "take their eyes off of the division of the surplus as the all-important matter, and together turn their attention toward increasing the size of the surplus until this surplus becomes so large that it is unnecessary to quarrel over how it shall be divided." Only recently have most Americans been led to confront the idea of limits to growth—and the limitations of the technologies we have chosen.

The first Europeans to settle in America brought with them that technology which had been one of the triumphs of the Middle Ages. The square-rigged sailing ship (its course set by the compass and sides bristling with cannon), the heavy plow, and the stirruped horse saddle, all made it possible for the newcomers to hold the native peoples at bay and wrest a living from the land. Mills were built to grind grain and saw lumber, full cloth and smelt iron, but most crafts were still carried on by hand and transportation was limited largely to horseback and boat.

Then, beginning with James Watt's improved steam engine of 1763, an astonishing series of new machines and processes began the Industrial Revolution in Great Britain which made much of the rest of the world's work obsolete. In quick order transportation (canals and turnpikes), the textile industry (the spinning jenny and power loom), and iron making (pit coal, the puddling furnace, and rolling mill), all combined with the rise of the new profession of civil engineer to create an industrial technology about which Americans knew little. First the shackles of British legislation designed to retard American industry, then the coming of war itself and lack of trans-Atlantic intercourse, kept the developing nation in ignorance.

With the winning of independence and the reestablishment of normal relations with Europe, the founding generation discovered that (as it seemed to them) a wise providence had provided them with a new technology capable of supporting their new nation. With abundant resources but few people to exploit them, Americans who aspired to surpass quickly the splendor and power of the old British Empire soon realized that machines would have to replace hands if the job were to be done. A patent law was passed in 1790, but more importantly, there was almost an explosion of what came to be called "Yankee ingenuity."

Typical of this startling burst of inventive activity was the automatic flour mill of Oliver Evans. In 1785 he put in operation a mill which moved grain and flour about by means of buckets on belts and endless screws. So fundamental was this innovation that it remained standard in American mills for nearly a century —until the introduction of the Hungarian (roller) process in 1870. At about the same time two inventors, John Fitch and James Rumsey, built and operated steamboats on the rivers of the East Coast. By the time of Robert Fulton's triumphant *Clermont* voyage from New York City up the Hudson River in 1807, over a dozen American inventors had built experimental steamboats.

One of the nation's most basic contributions to modern technology, the so-called "American System" of manufacture of interchangeable parts, was also one of its earliest. Despite precursors in

Sweden, France, and England, it was in the United States national armories (and reputedly in the private factory of Eli Whitney) that the use of machine tools to fabricate small arms from interchangeable parts was first placed on a practical and continuing basis. Within a few years the practice had spread to the manufacture of clocks, door locks, sewing machines, and similar devices made up of a large number of small metal parts. The dramatic lowering of costs through the American System made all of these devices available to a class of citizen who could not otherwise hope to enjoy them.

The sewing machine, patented by Elias Howe in 1846, was a key device on several counts: not only did it become one of the major technologies effecting the role and status of women, it greatly stimulated the ready-to-wear garment trade. In addition, it served to pass the technique of mass production on to such later important industries as bicycle manufacture, and through it in turn to the automobile industry. By 1860 some 110,000 sewing machines a year were being manufactured and sold.

Two other American devices of these years deserve particular mention because they reflected a primary need of the nation—that of dealing, somehow, with the great scale of the country. In 1834 Cyrus Hall McCormick patented his reaper, and at the first world's fair, the London Crystal Palace Exhibition of 1851, it was said of this device that it was "the most valuable contribution from abroad, to the stock of our previous knowledge, that we have yet discovered." By applying horse-power to the difficult task of cutting wheat, McCormick made possible the rapid growth of the great wheat producing regions of California and America's "Middle Border."

The vast distances which the United States presented dictated the direction of much of her inventive effort. McCormick's reaper was designed for large farms with few workers. The electric telegraph of Samuel F.B. Morse (first successfully operated in 1844) was another device designed to break down the isolation of the distant parts of the nation. It was, in addition, the first large-scale and commercially important use of electricity.

Along with the telegraph, the railroad was a major factor in linking together the American continent. Although an English invention, the steam railroad was quickly seized upon by Americans, improved in numerous ways, and used to tie together a national market for commerce. The completion of the first transcontinental line in 1869 was a triumph of both engineering skill and organizational management.

The needs of commerce and politics also encouraged the invention of a set of machines designed, in one way or another, to facilitate communications. The typewriter (1873) mechanized the task of correspondence and created a job category (that of secretary), which soon was opened to, and then dominated by, women—quickening their employment outside the home in growing numbers. The typesetter was closely allied to the typewriter, and, together with the steam-powered, "lightning" rotary presses of the same period, led to a golden age of inexpensive, popular newspapers. In 1876 Alexander Graham Bell, a researcher into the problems of the deaf, developed his telephone. It, too, was a business machine at first.

Important as these inventions were in building up the nation and cementing its economic and political unity, they were quite traditional in the sense that they depended in some degree upon the flash of genius and the heroic entrepreneur. The 19th century, however, was rapidly becoming one in which scientists were able to demonstrate a growing measure of control over natural forces. With new scientific competence, and confidence, came a set of institutions designed to stimulate, transmit, and apply this new technology. The U.S. Military Academy at West Point (1802) was the nation's first engineering school, but it was joined later by many private polytechnic institutes and public engineering schools. The first engineering society, the American Society of Civil Engineers, was established in 1852 and followed by the American Institute of Mining Engineers (1871), the American Society of Mechanical Engineers (1880), the American Institute of Electrical Engineers (1884), and others even more specialized. In these schools and societies the beginnings of scientific technology grew out of a living blend of experience and experiment. National legislation in 1863, during the American Civil War, led to the establishment of an agricultural and mechanical school in each state of the Union, and in that same year the National Academy of Sciences was chartered in the hope of providing the federal government with the best available scientific and technical expertise.

The new and characteristic technology of the 20th century came more from organized scientific research than had previous innovations, in part thanks to these new professional institutions. Thomas A. Edison had used a well equipped and staffed laboratory to produce such late-19th-century marvels as the incandescent electric lamp (1879), but the building of a corporate industrial research laboratory in 1901 by the General Electric Company (successor to Edison's own firm) marked the real beginnings of industrial research in the country. Organized by Willis R. Whitney, the laboratory was able to make rapid improvement in the efficiency of electric lamps and employed well over 300 people by 1918. So good was the science carried on in the laboratory that in 1932 Irving Langmuir was awarded the Nobel Prize for work done there in surface chemistry.

By 1917 the number of industrial research laboratories in the country had risen to 375, and in 1931 there were over 1600. Private industrial efforts were supported by large scientific efforts on the part of many agencies of the federal government—the Bureau of Mines (1910), the National Bureau of Standards (1901), and the National Advisory Committee on Aeronautics (1915), to mention only a few. In most states agricultural and engineering experiment stations (financed in part by the individual states, in part by the federal government) carried on economically important research, while universities attempted at the same time to create new knowledge, pass it on to students, and help society to apply it in a practical manner.

This institutionalization of innovation was matched by attempts to rationalize the other factors of production of American industry. Scientists and engineers, under the leadership of the forester Gifford Pinchot, instituted a conservation movement early in the century to put resource exploitation on a managed and efficient basis. This was particularly critical in terms of energy fuels. Since the birth of the American Republic reliance has shifted from renewable (and relatively non-polluting) energy sources such as wind, water, wood, and animals to non-renewable sources such as coal, oil, and natural gas. At the same time, total energy consumption increased approximately five fold from 1850 to 1900 and was to be eight times larger in 1975 than it had been in 1900. An increasing amount of this energy, of course, was consumed in the form of electricity. Networks of electrical power supply spread rapidly, especially after the federal government established the Rural Electrification Administration in 1935.

During the same period, sparked by Frederick Winslow Taylor's classic study *The Principles of Scientific Management* (1911), the concept of efficiency was applied to human-machine interactions in the factory environment. Just as the flash of inventive genius was replaced by planned scientific research, and the natural forests and free-flowing streams were replaced by managed tree farms and waterways controlled by engineering, so now were traditional craft skills replaced by scientifically defined and determined tasks.

The productive culmination of these various attempts at rationalization and scientific management was the assembly line, introduced into his automobile plant by Henry Ford in 1914. Based upon the principles of division of labor, the delivery of work to the workers, and the "planned orderly and continuous progression of the commodity through the shop," Fordism (as it came to be called) became not only the model for 20th-century industrial production, but the symbol, through Charlie Chaplin's film *Modern Times* (1936), of the dehumanization of industrial technology as well.

The entrance of the United States into World War II in 1941 led inevitably to a new federal concern with this, by now, large and productive scientific-technological capability. The Office of Scientific Research and Development (OSRD), established in 1940 and headed by Vannevar Bush, directed a massive effort to improve the implements of war and, as one historian has pointed out, succeeded in creating a totally new "electronic environment" for war. Such devices as radar (developed in Great Britain) and controlled nuclear reactions presented the postwar world with new technological potentials. The computer, systems theory, jet propulsion, and DDT were only a few of the marvels pushed forward by war. At the same time, the government moved to expand and enhance the capability of universities in basic science. The plan to place an astronaut on the Moon, announced in 1961 and accomplished in 1969, was a symbolic affirmation not only of the new technologies, but of a continued confidence in people's abilities to shape them to their purposes.

As the United States enters its third century, it is quite properly turning its attention to the costs as well as the benefits of its legacy of scientific and technological prowess. Recent awareness of the limits of growth, both in terms of reduced resources and increased pollution, has forced a closer look at both old and new technologies. The establishment of an Office of Technology Assessment in 1972 was only one indication of the realization that machines and processes are not always and in all ways beneficial. As the nation looks back to assess its past and forward to chart its future, it is certain that science and technology will play a crucial role in whatever future it chooses to pursue. They will be a science and technology more carefully tuned to human needs and limitations than previously, but confident still of their ability to serve the national purpose.

2

THE ARTISAN DURING AMERICA'S WOODEN AGE

Brooke Hindle

Early American technology depended primarily upon the skills of its artisans. Most of these skills were ages old, as were the basic tools of the trades. They were transmitted to America in the first instance by craftsmen who crossed the Atlantic, and they were perpetuated here by an apprenticeship system which taught skills by example. This process was supplemented by a continuing influx of European artisans as well as by books and manuals and occasionally by visits of Americans to Europe.

Skill was the great requirement in this pattern of technology. For example, anyone who had seen a barrel knew how the staves had to be curved and shaved to produce a tight vessel. The design was wholly without mystery. Yet, the skill of eye and hand required could not be quickly approximated by the uninitiated. The slightest error in the angle given the edge meant a leaky vessel. Similarly, the objective of the wheelwright in making the well-known wagon wheel was apparent to everyone. The fabrication of a satisfactory wheel, however, required practice and skill in making the parts and in putting them together.

High levels of skill were attained and practiced in a large variety of crafts carried on in colonial towns and cities. On the other hand, most of the early population was rural, and farming was the predominant occupation. Farmers were often separated by large distances from the skills of the cooper, the joiner, and the cabinetmaker—to say nothing of the silversmith or the gunsmith. In fact, even when they lived close enough to a town boasting such crafts, the price of the product was more than most farmers could afford. As a result, many in America were handy at a variety of handicrafts but skilled in very few. As best they could, they built their own barns and houses, made their own furniture and clothing, fabricated many of their own tools and utensils, and bought from practiced artisans only what they could not improvise.

In this setting, American technology did not differ strikingly from that of Europe, but in one respect, Americans enjoyed a great advantage. Especially by comparison with the mother country, Great Britain, Americans confronted a wonderful plenty of wood. The first colonists did not—as is sometimes imagined—find an entire continent covered by a climax forest. Even along the seaboard, the forest cover was broken at many points. Nevertheless, fine trees of all sorts abounded, and throughout the early period those who pushed forward the frontier continued to encounter new forests. By the end of the colonial period the price of wood had risen in the eastern cities, but, for most Americans, wood was plentiful in a measure unknown to contemporary Britons.

The ready availability of wood gave Americans advantages that have seldom been clearly appreciated. Wood was a basic foundation of the economy. Most obviously, it constituted the primary building and manufacturing material. Wooden houses and buildings of all sorts were used to a degree unknown in Britain or in the cities of Europe. Furniture, furnishings, boats, ships, wagons, carriages, and all manner of implements were made of wood. Secondly, wood was the primary fuel of this era. It was used for space heating, and for cooking, albeit inefficiently; it was used, sometimes in the form of charcoal, as the source of the heat required in all metallurgical and industrial processes. Thirdly, it was an important chemical product: in the form of potash, an industrial alkali; charcoal for gunpowder; and oak bark for the tannin needed in tanning leather. No other substance was as important in the technology of the time as wood.

The American supply of wood conferred many advantages, a few of which had certain negative aspects as well. Iron, strikingly, was one of the most important products of the wooden age. It was produced by heating iron ore with charcoal and a limestone flux in a furnace. Because England had been so denuded of her trees, she came to a point where she was unable to exploit her still rich iron mines but had to depend heavily on northern Europe for much-needed iron. Her American colonies had both iron ore and wood in plenty, with the result that iron production was encouraged and it became very successful. However, when England developed coke smelting, the colonies did not follow suit because they still had plenty of wood, and charcoal iron was a better product than coke iron. Unfortunately for the Americans, coke smelting in Britain soon led to other technological improvements and was integrally linked with the emergence of the Industrial Revolution. Early in the 19th century, Americans lagged in their industrial development precisely because their plenty of wood led them to cling, anachronistically, to charcoal iron.

In many applications of wood, the American advantage of plentiful supply seemed unalloyed. Shipbuilding, for example, was one of the most essential enterprises of the time. Wood was the basic raw material required, and England was much disadvantaged because she did not have the wood required to build the ships needed to maintain her extensive commerce. She came to depend upon America, which supplied not only masts and timber but American-built ships. America also built her own ships—for commerce, whaling, and fishing. By the time of the American Revolution, one-third of the ships flying the British flag were American built. American commerce and prosperity rested heavily upon the ships she had learned to build so well because she had so much available wood. But for her shipping and her iron, which she then produced in greater quantity than the mother country, America would never have been able to win her War of Independence.

The technology of shipbuilding was typical in the manner in which it rested upon individual skills. It depended upon a complex of arts or crafts which were learned by experience in an apprenticeship pattern. The development of extensive shipbuilding capabilities in the larger cities encouraged the growth of the spectrum of crafts which were required in completing a ship. Indeed, a large proportion of the crafts available in a seaport town, such as Newport, Rhode Island, had to be called upon to complete an oceangoing ship, the largest technological enterprise attempted in

The Flying Cloud, *one of the fastest of the wooden clipper ships built in 19th-century New England shipyards, could cover 433 statute miles in one day.*

Source: National Archives

the early period. Not only shipwrights, carpenters, and caulkers were required but also joiners, blacksmiths, whitesmiths, instrument makers, painters, glazers, ropewalks, and sailmakers. Many artisans who did not even make or work with wooden products were sustained by the great shipbuilding effort which rested squarely upon the supply of wood.

In the colonial era, when the craft skills upon which this Wooden-Age technology rested were being assimilated, American inventiveness was not notable. Changes occurred only in tools, techniques, and designs slowly, and many of the differences noted between European and American patterns were explainable by differing needs or differential supplies of materials or skills. In America, for example, many more structures were built of wood than of stone or brick as in Europe. In New England, especially, where building stone was lacking and good brick clay was scarce, wood construction was nearly universal except in the larger cities. Wooden structures served along most of the frontier. Tile roofs, introduced at points, quickly gave way almost everywhere, predominantly to wooden shingles or thatch but sometimes to slate.

Such changes and improvements as did emerge were sometimes silent or evolutionary, as in the case of the elegant American axe. The emergence of the American axe is undocumented and cannot be attributed to any known person or persons. It was the product of small changes and, no doubt, of much trial and error. It represented a remarkable improvement. The American axe was more compact than its European prototype; it was beautifully balanced so that the poll and the bit were nearly equal in mass and the center of gravity near the centerline of the handle. The handle was not a straight pole any longer but was gracefully curved—and indeed often fitted to the height and swing of the individual axman for whom it was made.

The result was a dramatic improvement in the speed with which one could fell trees. Trials demonstrated that an American with his axe could fell three times as many trees in a given time as someone with a European axe. The improvement cannot be attributed to higher levels of inventiveness in America but only to the great opportunity offered in the vast increase in the use of the axe in the New World.

A similar, but much broader-based technological achievement was recorded in the 19th century when the United States seized leadership in the development of woodworking machinery. Powered woodworking machinery had been a part of the American scene from a very early date. Sawmills were second only to gristmills in number during the colonial period; they depended primarily on waterpower and provided large quantities of sawed lumber not only for the colonies but also for export, especially to the West Indies. Again, the sheer quantity of lumber cut and used was the base upon which many American innovations were based.

Gang saws, muley saws, and other refinements upon the sawmill followed. Shapers, planers, and devices for finishing wood were developed further and faster in America than elsewhere. Some of these were specific products of individual inventiveness, as for example, the Blanchard lathe. Invented by Thomas Blanchard, it could make automatically a complex wooden form—following the pattern of a model. In this case, gunstocks could be produced much more quickly and with less skill on the part of the operator. The principle was quickly and widely extended.

Wooden balloon frame housing is quickly built and requires relatively simple skills as demonstrated by these Amish farmers working on a home in Indiana.

Another specific invention which grew out of the vast plenty of wood in America was the balloon frame house. This was introduced in 1833 by Augustine D. Taylor in Chicago. Instead of the heavy timbers which had to be carefully fitted and joined in preceding house framing, the balloon frame was made by using many lengths of light lumber—ultimately by two-by-four-inch boards rather than by ten-inch square timbers. This effected great economies in the amount of wood used. It also speeded the job of framing a house and, again, reduced the skill required. A balloon frame house was made by simply nailing the pieces in—eliminating the laborious process of cutting mortise and tenon joints or drilling and fitting trenails.

This procedure of mechanizing processes and simplifying the work remaining to be done by the artisan was very directly encouraged in lumbering and woodworking. American advances in this direction were notable because of the plenty of wood and the large role played in the economy by wood and wood products.

At the same time, other forces also encouraged mechanization and simplification. Americans received very rapidly the advances of the Industrial Revolution, which coincided in some of its key beginnings with the era of the American Revolution. They were especially receptive to the mechanization of textile manufacturing and to the introduction of the steam engine. Indeed, they were developing their own significant inventions and innovations—in steam engine construction by the first decade of the 19th century and in textile machinery by the 1820's. Then, on their own, the

Americans pioneered in the development of interchangeable parts in combination with the factory system to open the access to mass production in what the British called "The American System of Manufacturing."

The great advances in woodworking machinery were achieved by applying the developing techniques of mechanization to this great area of need—wood cutting and shaping. The plenty of wood set the stage for one of the most striking applications of the "American System"—to clockmaking. The use of interchangeable parts had arisen in gun manufacturing and one of its primary areas of development was the Connecticut River Valley. Here, Eli Terry made a similar move from handmade brass clockworks to wooden works in which the wheels were cut to standard patterns in large quantities. These wooden, mass-produced clocks, because they were so cheap, rapidly became dominant in the market. The next step, taken most successfully by Chauncey Jerome, was to make brass clockworks on the basis of interchangeable parts—a more difficult process which would have been retarded if the transition through the easier construction of mass-produced wooden works had not occurred.

Blanchard, Taylor, Terry, and others were artisans or craftsmen who proceeded in much the same manner artisans had for centuries. They learned a trade which they practiced successfully, beginning with the tools and techniques commonly used in their time. In their initial work, they copied familiar models—the age-old method and objective.

What led them to break out of established patterns of production to invent new, simplified, mechanized ways for making products —and to change the products in the process? The opportunities in this period of history, and especially in America in wood-based technologies, encouraged them to invention and innovation. The strongest line of encouragement led to simplification of process and product and to mechanization. This was, indeed, the same sort of encouragement which had led a little earlier in Britain to the classical Industrial Revolution and to achievements which had great appeal and influence in America.

Yet the encouragement to invention and innovation did not always lead to mechanization. In the wood-based technologies, a more ancient sort of change is exemplified in the construction of truss bridges, at first made entirely of wood. Such bridges were used in road construction for crossing streams, but American road construction long remained underdeveloped and inferior. The big need for cheap truss bridges—and the strong monetary encouragement to invention and improvement—arose with the great railroad building effort. American railroads, covering vast stretches of geography, had to be built as cheaply as possible. Tight curves and steep slopes were accepted and cheap bridges were preferred to expensive cutting and filling or route detours.

The result was to encourage inventors to invent new truss forms that would be strong enough and cheaper than those being built. The objective was to use less material and, secondly, to use less labor. For a time a great number of patents were issued for new trusses, each inventor hoping that his invention would be widely adopted and make him rich. New and better trusses were devised in a process of simplification and economy of construction—but not of mechanization.

In another wood-based technology, shipbuilding, American encouragement to change and improvement also failed to produce mechanization but, as in the case of truss bridges, spurred what should be called design improvement. Americans had much experience in the construction of wooden ships and of wooden boats for inland transportation. In each case, specifically American developments were often keyed to specific needs. For example, the Durham boat, a large, low-draft craft for carrying bulk cargoes, was used on the eastern rivers. The western keelboat had similar missions. The Baltimore clipper was a specifically American sailing craft, well designed for the coastal sailing intended, and, in the later international clipper ship competition, the Americans registered significant design achievements.

Perhaps the kind of change encouraged in shipbuilding is best epitomized in the design of the United States frigates begun at the very end of the 18th century. The overall concept of these ships was that of the leading American shipbuilder, Joshua Humphreys, who produced something different from either the French or British prototypes. His 44-gun frigates were larger and more heavily armed than similarly rated ships, and they had fine lines which made them better sailers. They were creative innovations in the classical sense of representing improvements over existing models—rather than differences in kind.

Artisans, then, came out with many kinds of change, invention, and improvement based upon America's wonderful plenty of wood in combination with the variety of needs and opportunities which arose. They produced a great number of striking design improvements keyed to local circumstances. Many of their innovations did contribute to the 19th-century movement toward simplification and mechanization. Invention was heavily encouraged, and American artisans were often led to think more about change and improved production than about excellence of finish or craftmanship.

Certainly many American characteristics were fed by this process. It has been suggested, for example, that the "replacement economy"—that is, the manufacture of cheap goods which could be replaced after a short period of use instead of being repaired or used in a worn state—may have grown out of the experience with wood-based technologies. Also, because wood was plentiful, the Americans did not think much about waste. They used saws that ran faster and cut more efficiently than any predecessors— unconcerned that they wasted more in sawdust than other saws.

The experiences with wood changed America, changed the nature of many technologies, and changed the world. They changed too the craftsmen who were themselves the agents of these changes. Today more than before, we are aware that the improvements were purchased at the cost of some real disadvantages. All changes were not for the better ultimately. In their day, however, they represented great achievements in which the negatives were relatively unimportant. They played a major role in leading us to our present world.

3

THOMAS JEFFERSON AND
A DEMOCRATIC TECHNOLOGY

Hugo A. Meier

The election of Thomas Jefferson as the third President of the new United States of America in a so-called "Revolution of 1800" marked the fulfillment of the republican ideal in the Western world. Embodied in that achievement was a tacit recognition of the American faith in people's capacity to wed a political ideology of freedom for all people with the practical goal of uniting advances in knowledge to the well-being of the ordinary citizen.

Thomas Jefferson, republican ideologue, sophisticated citizen of the late 18th-century world of science and letters, remains for us, in our technology-dominated 20th-century, America's earliest distinguished spokesman for a democratic technology.

Jefferson's own Virginia background did not seem propitious for what we in the 20th century conceive to be the democratic society. Here existed a close-knit, elitist culture rooted in the conservative traditions of an agrarian-based social order, itself strongly dependent upon human slavery. But America in 1776 was still too new a land to have fostered any genuine aristocracy. Indeed, Peter Jefferson, Thomas' father, had been obliged to make his own way as surveyor and succeeded—as so many of the Virginia gentry were to do—by the shrewd acquisition of baronial acres bargained out of the vast reserves of forest and farm lands which the New World afforded.

America was then already the land of opportunity for those with initiative. Judicious marriages with established families and their generous landholdings provided the obligations, incentive, and means by which an excellent education and experience in public affairs might be acquired and a responsible sense of paternalism developed. Thomas Jefferson shared this patrimony of landed wealth and respect for knowledge. He, as had George Washington, acquired the surveyor's skills. By inheritance and by assiduous accretion of acres, Jefferson's lovely estate on the Rivanna River, Monticello, reached an impressive size. Its master joined not only the 18th-century world of refinement and culture but participated avidly in its intellectual pursuits.

For Jefferson the scholar and political ideologue, Europe and in particular France constituted a congenial spiritual home. Appointed as Benjamin Franklin's successor as minister to the Court of Louis XVI, Jefferson established an easy rapport with French men of science and letters. He was less happy with England and the English culture despite his own intellectual debts to both. The burgeoning Industrial Revolution in England, however, left Jefferson with considerable admiration of British technology and a disenchantment with its less happy urban and human consequences. He was keenly aware that the focus on experimental science, which fascinated French philosophers, in England was directed at converting labor and resources into useful goods for everyone.

At home, the colonies sought and found their freedom in the midst of embarrassingly rich natural resources. These subtly shaped popular tastes and entrepreneurial talents toward satisfying domestic wants. Science *per se* was left to the preoccupation of leisured clergymen and other amateurs, or confined to the lecture rooms of America's embryo colleges. Technology was for most colonials the natural response to the practical demands of their environment. Less of the "know-what" and more of the "know-how" approach to knowledge fell quite naturally within the interests of the pragmatic Americans. They could borrow freely from the knowledge and expertise of Europe—a function which Jefferson also served via his tours abroad, his beloved books, and the streams of knowledgeable visitors who climbed the slopes to Monticello.

As a member and a president of the American Philosophical Society, Jefferson acted as intermediary between its members and the foremost European men of science. His vast correspondence addressed to knowledgeable men on both sides of the Atlantic reflected in its diversity and devotion to learning the spirit of the volumes of Diderot's illustrious *Encyclopédie,* copies of which Jefferson sought out for grateful American colleagues. His own library, despite losses by fire, represented a cross-section of Enlightenment knowledge. When British armies burned the national Capitol at Washington in 1814 and its library of perhaps 3,000 volumes, Jefferson contributed eleven wagonloads of volumes from his own library—about 6,500 books—as the nucleus of a new Library of Congress.

Such devotion to learning was not uncommon even amidst the war-harried and empirically minded scholars of the New World. That a people traditionally practical and devoutly religious might also have reservations about the strange social and political dogmas of the French Enlightenment was also apparent. They had lost enthusiasm for continuing revolutions in a renewed quest for social stability and economic success, and Jefferson's affinity for French revolutionary philosophies and his religious skepticism encouraged the beliefs of his Federalist Party opponents of his singular unfitness for public office.

As a man of the Enlightenment, Jefferson was not content with accepting the world unquestioned. The universe itself now was a

Thomas Jefferson (Painting by Rembrandt Peale)
Source: New York Historical Society

solvable problem—Newton had seen to that. If Jefferson was intensely curious about that universe, its laws, its celestial mechanics, its magnificent global Earth—that curiosity was aimed at understanding its mysteries in ways which might lead to the improvement of the human condition. "Perfect happiness, I believe, was never intended by the Deity to be the lot of one of his creatures in this world," he observed in 1763, "but that he has very much put in our power the nearness of our approach to it, is what I have steadfastly believed."

The 18th century with its faith in science and reason was appropriately the century of republicanism as a favored political expression of both concepts. "Science," said Jefferson, "is more important in a republican than in any other government." Not only was a republican environment especially congenial to science and technology, he believed, but the combination contributed to the cultivating of a new sense of American nationalism. Granted, as Jefferson had said, that "in an infant country like ours we must depend for improvement on the science of other countries, longer established, possessing better means, and more advanced than we are," the New World offered unparalleled opportunity for scientific and technological discovery. "It is the work to which the young men who you are forming, should lay their hands," he advised President Joseph Willard of Harvard College. "We have spent the prime of our lives in procuring them the precious blessing of liberty. Let them spend theirs in showing that it is the great parent of *science* and of virtue; and that a nation will be great in both, always in proportion as it is free."

The new national pride of the versatile scholar of Monticello kept him wary of foreign claims to inventions which he believed to be of American origin—and especially so where the patronizing Mother Country was involved. The new Patent Office was for Jefferson a symbol of American creativity, and he confidently advised William L. Thornton, its administrator, "I think the English must acknowledge we excell them in the faculty of invention. There can be no better proofs than are furnished by our office." He shared a common American suspicion of the employment of European engineers and technicians by the new government because the practice would inhibit the development of native American talents to the detriment of a truly national technology.

Jefferson's sense of American nationalism sometimes goaded

him into exaggerated rebuttals of European criticisms of the poverty of American achievement in science and invention and unacknowledged borrowings from abroad. He pointed out that by comparison of populations France might be expected to show six geniuses and England at least three for every American genius. But he also believed in the internationalism of science, refused to restrict in any way foreign use of his own inventions, and urged that nations reorganize their international relations with the same harmoniousness characteristic of scientific societies, which constituted "a great fraternity spreading over the whole earth."

If Jefferson believed science and technology to be important to the success of a republican form of government, he was even more concerned with their impact on the everyday life of the citizens of that republic. Although the term democracy was not on the popular tongue of Americans during most of Jefferson's life, his faith in the common man found expression in his concern with applied science and social change. In a democracy flexibility and change were conditions most favorable to technological advancement. Reason, so well exemplified in scientific progress, must replace superstition. Knowledge must replace ignorance. Conservatism must accommodate to receptiveness toward social change. He accused the New England clergy of insisting on looking back too exclusively to the opinions and practices of their forefathers "instead of looking forward for improvement." To Robert Fulton, recently acclaimed for his successful steamboat, Jefferson wrote in 1810: "I am not afraid of new inventions or improvements, nor bigoted to the practices of our forefathers. It is that bigotry which keeps the Indians in a state of barbarism in the midst of the arts."

If reason and talent were to be at the service of every man, then utility should be the final judge of merit in a democratic technology. This capacity to serve the needs of the average citizen also served the highest humanitarian goals. The line between pure science and technology was blurred in the new republic: a qualified scientist such as the mathematician-astronomer David Rittenhouse felt sometimes burdened, but seldom offended by his assignment to duties of clock-making or surveying. Jefferson planned to introduce branches of technologically oriented education into the curriculum of his proposed University of Virginia, observing that "I have taken some pains to ascertain those branches which men of sense, as well as of science, deem worthy of cultivation." Existing schools, he complained, taught their versions of science "too much *in extenso,* for the limited wants of the artificer or practical man."

In his concern for applying science to useful ends, Jefferson invites inevitable comparison with Benjamin Franklin. Both men shared the Enlightenment curiosity about man and his world. Both favored a practical view of science. Jefferson shared the point of view which Franklin expressed in advice to a correspondent in 1783 who was "endowed with a genius for the study of nature." Franklin wrote, "I would recommend it to you to employ your time rather in making experiments, than in making hypotheses and forming imaginary systems, which we are all apt to please ourselves with, till some experiment comes and unluckily destroys them." Franklin drew from *Poor Richard* his example of the ingenious striking sundial which, by action of the sun on twelve burning glasses and trains of gunpowder, would fire off 78 guns in proper sequence: "Let all such learn that many a private and many a public Project, are like this Striking Dial, great Cost for little Profit."

Although admitting himself to be an empiric in natural philosophy—"suffering my faith to go no further than my facts"—Jefferson did admit the usefulness of hypothetical speculations "because by the collisions of different hypotheses, truth may be elicited and science advanced in the end." Franklin proved to be the better scientist—a close observer and experimenter in natural phenomena as well as the practically minded inventor of improved fireplaces. Jefferson hoped to use his retirement years for the sustained study of natural science, but in his advanced years he recognized that science had progressed considerably beyond his capacity to catch up with it.

His pragmatic concern for using science to enhance human comforts made Thomas Jefferson a close and curious observer of things useful. Friends and acquaintances drew freely upon his generosity while he was abroad to obtain for them those things helpful to the citizens of his republic. His self-professed "zeal to promote the general good of mankind by an interchange of useful things" resulted in a stream of ideas, books, commodities, and inventions likely to improve American technological know-how. What might appear to be mere gadgets to some—improved letter-presses, efficient cylinder lamps burning olive oil—to Jefferson were instruments serving the inquiring mind. He found matches to be "a beautiful discovery, and very useful" in such humble tasks as lighting candles, kindling fires, and sealing letters.

The dedication of ingenuity supported by a "practical" science to domestic purposes was staunchly defended by Jefferson. It represented for him the democratic goal of a useful science, however humble those uses. "You know the just esteem which attached itself to Dr. Franklin's science, because he always endeavored to direct it to something useful in private life," he advised Dr. Thomas Cooper at Carlisle (now Dickinson) College, who was planning a course of instruction in chemistry in 1812. "The chemists have not been attentive enough to this. I wished to see their science applied to domestic objects, to smelting, for instance, brewing, making cider, to fermentation and distillation generally, to the making of bread, butter, cheese, soap, to the incubation of eggs, etc. And I am happy to see some of these titles in the syllabus of your lecture. I hope you will make the chemistry of these subjects intelligible to our good housewives."

He approved the new steam engine of one George Fleming in 1815 because it was simple, cheap, and adaptable to the most common purposes. "A smaller agent, applicable to our daily concerns, is infinitely more valuable than the greatest which can be used only for great objects. For these interest the few alone, the former the many." He advised him to develop "a domestic machine" to be used for lifting water, washing linens, kneading bread, heating the hominy, churning butter, and turning the spit. If these seemed trivial, he argued, "of how much more value would this be to ordinary life than Watt's and Bolton's [sic] thirty pair of millstones to be turned by one engine." A little self-conscious about his improvements of the common plow, Jefferson felt obliged to apologize to Sir John Sinclair, President of the National Board of Agriculture at London in March of 1798, that "the combination of a theory which may satisfy the learned, with a practice intelligible to the most unlettered labourer, will be acceptable to the two most useful classes in society."

But behind utility lay Thomas Jefferson's wide-ranging curiosity and richly eclectic knowledge. Monticello's halls and walls, its very architectural design, its kitchens, workshops, and household furnishings, all reflected the diversity of his interests. His library was rumored to contain multi-volumed studies of prairie-dog genealogy and of the agility of the flea. He was as much interested in the physical and acoustical means by which his fiddle produced music as in the music itself. A consuming interest in mathematics persuaded him to decimalize the new republic's currency and to reform its inherited archaic system of weights and measures, "bringing the calculations of the principal affairs of life within the arithmetic of every man who can multiply and divide plain numbers."

As citizen of a militarily weak nation hovering behind an extensive sea coast and enjoying fine harbors, Jefferson shared a prevailing defensive stance in foreign policy. Thus he welcomed Robert Fulton's invention of "torpedoes" or underwater mines, observing that "I sincerely wish the torpedo may go the whole length you expect of putting down navies." Jefferson, however, also supported the idea of an efficient American navy capable of defending harbors and consisting mostly of inexpensive gunboats. At least 176 such vessels were constructed during his presidency—each ranging from 50 to 70 feet in length, visually unimpressive in light of the splendid vessels of His Majesty's navies, and nicknamed by his critics as "Jefferson's democratic sinking fund." He established a corps of engineers in 1802 as the nucleus of West Point Military Academy, favoring technicians over military men. And Jefferson shared the contemporary fascination with balloons which he believed, as did Franklin, might make war too hazardous for aggressors by exposing their own territories to aeronautical invasion.

Jefferson's ready ear for reports of novel discoveries and unique inventions led to the apocryphal story that he himself had invented an elevated saw mill operated by the power of wind on sails, that he had one constructed with theoretically good judgment on a hilltop, but had failed to consider the more practical problem of getting the heavy saw logs up to it. However, the master of Monticello was concerned with more than visionary sawmills. As Vice-President he despatched two emissaries to Europe for the purpose of reporting "Objects of Attention for Americans" in the field of European technology. Everything of interest to agriculture must be closely observed, as well as developments in the "mechanical arts so far as they reflect things necessary in America, and inconvenient to be transported thither ready-made, such as forges, stoves, quarries, boats, bridges." But as regarded the lighter mechanical arts and manufactures, some only might be worth a superficial view because "circumstances rendering it impossible that America should become a manufacturing country during the time of any man now living, it would be a waste of attention to examine these minutely."

That judgment, to be sure, was shared by the majority of agrarian-minded Americans. But while in France, Jefferson was fascinated in 1789 by early experiments with interchangeable parts and sent samples and instructions to General Henry Knox, the Secretary of War, with a vain hope that the inventor and his workmen might remove to the United States. In England he interviewed Matthew Boulton about the comparative efficiency of his steam engines and arrived at the impressive judgment that "a peck and a half of coal perform exactly as much as a horse in one day can perform." Meanwhile he continued to collect and ship to America new books of great variety in which European know-how and ideas could be made more widely available to Americans.

Jefferson foreshadowed the later world-wide reputation of Americans as lovers of mechanical gadgets. A faithful and perhaps obsessive correspondent, he was delighted with the "polygraph," a unique copying instrument invented by a Philadelphian. It consisted of a desk above which was attached a mechanical movement to which from three to five pens were suspended. He was exceedingly pleased with his new ability to write several "original" copies of letters while the device faithfully imitated his own pen strokes in making duplicates. His urge for perfection led Jefferson to attempt improvements, observing that "it gave me so much satisfaction that I thought it worth while to bestow some time in contriving one entirely suited to my own convenience." He called it his "portable secretary" and "the finest invention of the present age"—a notion not necessarily shared by Americans of a lesser epistolary enthusiasm, for the device did not become widely popular.

Agriculture was still the overwhelming preoccupation as well as occupation of most Americans, and Jefferson proposed to elevate it to a science so that all Americans could enjoy the benefits of contemporary experimentation in soil improvement, diversified farming, conservation, and plowing and harvesting techniques. While still Secretary of State, Jefferson heard of a new threshing machine on a farm near Philadelphia and went to look at it personally. Coming upon a description of a British harvesting machine,

Jefferson obtained a model and duplicated it at Monticello. Not fully satisfied, he constantly sought its improvement.

Most notable, of course, was Jefferson's own careful designing of a plow which, based on mathematical principles, was intended to minimize earth friction and thus conserve horsepower and actually plow several inches deeper with the same expenditure of energy. When in France in 1788 he had noticed the clumsiness of moldboards on plows being employed by peasants near Nancy. He computed the most satisfactory shape into which such an implement might be hewed, and later used such a plow at Monticello where he was pleased to observe that ordinary plows tended to wear into a comparable shape—practical evidence of its functional success. Jefferson did not patent the invention but was satisfied with public recognition of his achievement, including a gold medal from a French agricultural society and an honorary membership on the English Board of Agriculture.

The desire for self-sufficiency at Monticello interested Jefferson in any process or device which would contribute to that goal. Being responsible, as Secretary of State, for registering patents, he received Eli Whitney's required drawings of his cotton gin invention in October 1793, and hastened off a letter to Whitney filled with curiosity about that surprising machine. "One of our great embarrassments is the clearing the cotton of the seed," he explained. "I feel considerable interest in the success of your invention, for family use." He wondered if the machine had been tried

Jefferson's home, Monticello, which he designed and built himself near Charlottesville, Virginia, beginning in 1769, serves as a showcase for his practical ideas and inventions.

thoroughly or was "but a machine of theory"—another hint of Jefferson's concern for utility as the standard of invention.

Household technology offered Jefferson possibly his greatest opportunity to demonstrate his natural ingenuity. Visitors to Monticello have for nearly 200 years marveled at and often found amusement in the strange and clever and almost always utilitarian devices sprung from the mind—and frequently contrived by the hands—of the disciple of democracy. The list was long and various, and included a basic workshop and its tools. A turnstile-like device provided the convenient hanging and retrieval of coats and other clothing. A swivel chair permitted Monticello's master to obtain a circumambient view of his library-study. Among his adjustable music stands was one also convertible to a table. A walking-stick was expandable into three legs and a cloth seat to provide the convenience of a stool. A special mechanism opened double glass doors simultaneously.

There were other fruits of Jefferson's inventive imagination. The famous Monticello clock presented one face inside, the other outside the house, used cannon balls as weights, needed winding but once weekly, and not only struck the hours but also indicated the days. A clever weather vane over the east portico, connected by a rod to a dial in the ceiling below, informed the master as well as the neighbors about the direction of the wind. The ceilings at Monticello, too, were fire-proofed. Ever the restless wanderer, Jefferson utilized a pedometer to measure his walking tours and added an odometer to his carriage to count off the miles of more distant journeys.

Jefferson's pleasure in acting as host to many and often distinguished guests inspired an intriguing "serving panel" which expedited the serving of warm food to the dining room from the pantry. A central vertical swivel turned a single-panel door on whose shelves serving dishes might be placed to minimize the intrusion of the traffic and curious ears of servants in the dining area. Dumb-waiters, concealed behind paneling on either side of the mantel-piece, brought Jefferson's cherished imported wines from the cellar to the dining room.

While President and deeply concerned with national defense, Jefferson modeled a drydock intended to preserve his precious gunboats from storm and rot. Most of them would be kept in reserve as an early version of the "mothball fleet" idea following

27

World War II—defense coupled with economy. He proudly kept a model on his mantelpiece despite the tart reception of the concept by his critics.

A major issue confronting the new government was determining the proper kind of support for science and invention. The new Constitution was to provide for protection of invention by a monopoly patent grant for a "limited" number of years. Jefferson had strong but not inflexible notions about such support. Fundamental to his thinking was a belief in the freedom of ideas and the dangers inherent in restricting their applicability. In a letter to James Madison from France in 1788, Jefferson earnestly advised prohibition of even limited monopolies. Madison attempted to convince Jefferson that "monopolies are sacrifices of the many to the few," and since in America power lay with the many, there was little danger that the few would be unduly favored. Jefferson eventually conceded that progress in the sciences and arts would be promoted by securing temporary exclusive rights for original writings and discoveries.

Central to Jefferson's thought was the thesis that ideas were universal and "should freely spread from one to another over the globe, for the moral instruction of man, and improvement of his condition." Nature had intended ideas to be "expandable over all space" like fire and as "incapable of confinement or exclusive appropriation" as was the air men breathed. Moreover, once publicly disclosed, a new idea forced itself into the possession of everyone, and the receiver could not dispossess himself of it. "Its peculiar character too," said Jefferson, "is that no one possesses the less because every other possesses the whole of it. He who receives an idea from me receives instruction himself without lessening mine, as he who lights his taper at mine receives light without darkening me." Consequently, Jefferson conceived government's role to be that of disseminating information useful to the citizens rather than insuring material profit for innovators.

Invention, he also argued, was an accumulative process rather than the single fruit of genius. "The fact is that one new idea leads to another, that to a third, and so on through a course of time until some one with whom no one of those ideas was original, combines all together and produces what is justly called a new invention." Such thinking prompted his annoyance with British claims to what were to Jefferson at least equally American inventions. "I see by the Journal of this morning, that they are robbing us of another of our inventions to give it to the English," he protested in 1787. The "invention" consisted of a method for making a wheel of one piece which, Jefferson claimed, Philadelphians enjoying Sunday picnics across the Delaware River saw on every farmer's cart. The idea, brought to Europe by Benjamin Franklin, could be found in Homer—and that, said Jefferson, was where New Jersey farmers probably picked it up "because ours are the only farmers who can read Homer."

Troubled as he was by the demands of inventors under the new patent law of 1790 for grants of monopoly, Jefferson saw as great problems for technological progress under such a system as he did for the likelihood of the encouragement of innovation. Especially irritating were the sweeping claims of Oliver Evans whose genius in invention was coupled with a strong sense of property rights. Evans's automated gristmill—capable of grinding, sifting, and bagging flour by means of water power—was the object of mingled envy and resentment by millers generally. Jefferson had completed his own flour mill in 1806 with such improvements as elevators, hopper boys, and conveyor buckets. Forthwith, he was presented a bill for his use of these Evans-claimed ideas—and paid it, but not without some philosophical grumbling: "If the bringing together under the same roof various useful things before known . . . entitles him [Evans] to an exclusive use of all these, either separately or combined, every utensil of life might be taken from us by a patent."

Jefferson also strongly opposed attempts to monopolize the principles underlying classes of inventions and disputed Oliver Evans's claims on this basis. "I can conceive how a machine may improve the manufacturing of flour: but not how a *principle* abstracted from any machine can do it." He warned that such practice might eventually result in anyone who applied an axe, hoe, or spade to a new purpose demanding the exclusion by patent of similar use of these essential tools. Indeed, "if a new application of our old machines be a ground of monopoly," he observed, "the patent law will take from us much more than it will give . . . . A man has the right to use his knife to cut his meat, a fork to hold it; may a patentee take from him the right to combine their use on the same subject? Such a law, instead of enlarging our conveniences, as was intended, would most fearfully abridge them, and crowd us by monopolies out of the use of the things we have." The spinning machine, thus, should legitimately be usable for any purpose to which it was adaptable. A Georgetown smith, said Jefferson, had some 16 years earlier made a device to crush plaster which, in a modified form, was now an Evans monopoly for crushing corncobs.

Jefferson was also unhappy about the practice of permitting sales of patented devices by geographical districts only. "Every man," he said, "should be free to use anywhere what he can lawfully buy anywhere. This abuse with the plagiarisms committed & [sic] imposed on us render the advantage of the patent law problematical."

The length of such patent monopolies also concerned Jefferson, who in 1807 wrote Evans that while an inventor deserved the benefits of an invention for a certain time, "it is equally certain it ought not to be perpetual: for to embarrass society with monopolies for every utensil existing, and in all the details of life, would be more injurious to them than had the supposed inventors never existed; because the natural understanding of its members would have suggested the same things or others as good." He did admit, however, that the length of time for a patent might have to be conditioned by the American environment whose sparse settlement and vast distances naturally delayed the knowledge and spread of an invention to the disadvantage of a patentee.

Whatever may have been Jefferson's position philosophically on the touchy issue of patent monopoly, as Washington's Secretary of State he was confronted with the practicalities of applying the republic's first patent law. The act, passed in 1790, was to be administered by a board consisting of the Secretary of War, the Attorney General, and the Secretary of State. Jefferson's other duties, to be sure, already included foreign affairs, the preserving of federal records, and promulgating of the laws, and all to be accomplished by a very modest budget and small staff. But recommendations that the board consist of experts were stilled by prevailing fears of a new bureaucracy indifferent toward inventors' rights.

It was Thomas Jefferson, therefore, who was principally responsible for applying the new act—not unexpectedly in light of his reputation. His dislike for monopolies was now coupled with a concern for detail and accuracy, and a determination upon stan-

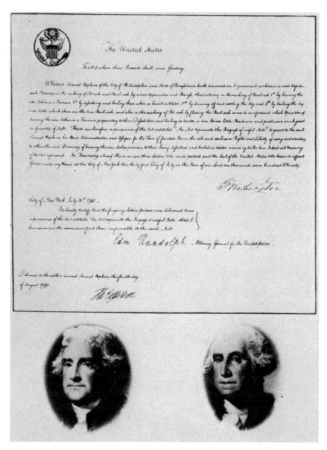

A copy of the first U.S. patent, issued in 1790, with two of the signers pictured: George Washington and Thomas Jefferson. This patent was valid for 14 years, but under a later law 17-year patents were established.

dards for invention based on their novelty, utility, and social importance. On July 31, 1790, the first patent was issued to one Samuel Hopkins for a new manner of preparing pot and pearl ashes. Only two other patents were issued in 1790. Meanwhile, Jefferson enjoyed his opportunities to meet and converse with men of scientific and technological interests.

The process of issuing patents was tedious. The Board of Arts met monthly to compare notes on applications, requesting proper specifications, drafts, or models. The complex task of determining credit for the new steamboat inventions resulted in four claimants receiving patents. Frivolous and inoperable devices were rejected. "Actual experiment must be required," insisted Jefferson, "before we can cease to doubt whether the inventor is not deceived by some false or imperfect view of the subject."

A classic example of Jefferson's cautiousness and thoroughness as patent administrator related to the application of Jacob Isaacs for a patent on a still and furnace designed to distill fresh water from sea water. In an era of sailing vessels and long, uncertain voyages such a device was scarcely unimportant. Jefferson required Isaacs to assemble his contrivance in the Secretary of State's office for a test. But he also called in David Rittenhouse, astronomer-clockmaker-surveyor, and Caspar Wistar, one of America's earliest glass technologists, to help judge the merits of Isaacs' method. After five tests the latter's request for a patent was denied. Jefferson's own long report pointed out how the idea was not a sufficient improvement over already known methods of converting seawater.

If his interests seemed universal, Jefferson's time and energy were not limitless. He soon felt oppressed by the mounting duties of his several offices, and his conscience was troubled by his doubtful capacity to judge fairly the increasing variety of ideas and inventions which the republic's more fertile innovative minds paraded before him. He reported that he felt "oppressed beyond measure by the circumstances under which he has been obliged to give undue and uninformed opinions on rights often valuable and always deemed so by the authors."

In February 1793, a new patent act greatly simplified the tasks of the Secretary of State's office by placing responsibility for judging the validity of issued patents in the district courts. While this move also met the democratic suspicion of government bureaucracies, it additionally resulted in so casual a process of registration of patents that it was claimed individuals copied and re-patented ideas from existing models in the patent office. Jefferson, happy to be relieved of the duties of search and examination, hoped that the new approach might lead to a new form of jurisprudence.

During Thomas Jefferson's term as administrator only 67 patents were issued—a sign less of the failure of American ingenuity than it was of the exacting standards of one who suspected monopoly of any kind and who possessed a mind sufficiently cautious and critical to make the patent right a privilege to be achieved only by genuine merit. The basic revision of the American patent system in 1836 resulted in a compromise by including Jefferson's rigorous tactics in a new system of examination to replace the largely clerical system succeeding him.

Jefferson's ideal of a technology dedicated to the popular weal both reflected and enforced an attitude which would increasingly dominate the American culture. The 19th century—and Jefferson lived through its first quarter—carried on that ideal in an ever proliferating abundance of practical inventions for everyday uses.

Alexis de Tocqueville, the French liberal politician and social commentator who visited the United States during the 1830's, found the Jeffersonian ideal thriving in a manner that might have disturbed even Jefferson. A worship of practical invention, a concern for utility, an enthusiasm for the material consequences of applied science—all, Tocqueville observed in his still cogent portrayal of *Democracy in America,* were the democratic fruits of science applied to the needs of the people. "Those who cultivate the sciences amongst a democratic people," he concluded, "are always afraid of losing their way in visionary speculation. They mistrust systems; they adhere closely to facts and the study of facts with their own senses."

The results of such materialistic application were to include a rising standard of living that would remain the envy of the world into our own times. Today while we fear Americans may be nearing the end of a long era of technologically created abundance, nonetheless we join with much of the rest of humankind in lamenting its consequences of environmental despoliation, unimaginable horrors of technically perfected warfare, and the depersonalizing of life itself.

There is, however, still present the historical consciousness that the Jeffersonian ideal of world cooperation and understanding via science and technology offers hope for humankind in husbanding its remaining natural and human resources and directing their benefits to the satisfying of the needs of everyone, wisely and humanely.

4

BENJAMIN HENRY LATROBE AND THE TRANSFER OF TECHNOLOGY

Darwin H. Stapleton

The term "transfer of technology" usually means the taking of a technical skill from a nation where it is already established to another nation where it is unknown. Seldom has a transfer of technology been accomplished except by the actual migration, temporary or permanent, of one or more skilled persons. Today, with multiple and complex means of impersonal communication and rapid international travel, this element of the transfer of technology may not be readily apparent, but transfers still depend largely upon the migration of skilled persons.

Undoubtedly the most critical era in the transfer of technology to the United States was the period extending from about 1790 to 1850. By the immigration of skilled Europeans and as a result of trips abroad by her own citizens, America kept abreast of the rapid industrialization of Britain and Western Europe. Moreover, by the judicious modification of what was imported and the substantial contributions of native mechanics and crafts people, the United States created its own technological tradition, which was soon admired throughout the world. For an understanding of the ways and means of the transfer of technology to America we must look at the lives of those who participated in the process. One of the most fascinating and significant persons was Benjamin Henry

Latrobe, who helped establish the professions of architecture and engineering in America.

Born in Yorkshire in 1764, Latrobe was educated in Moravian schools in England and then Germany. He showed an early aptitude for drawing and soon took an interest in the allied fields of architecture and engineering. After studying German engineering informally and touring the architectural monuments of Western Europe, Latrobe returned to England in 1783. There he probably spent a year or two in the office of John Smeaton, the foremost English engineer, and afterward he worked under Samuel Cockerell, a distinguished architect. In 1791 Latrobe established himself as an independent architect and engineer in London, and he thus became a member of two relatively small professions which scarcely existed in the United States.

One of the aspects of English engineering which at this time differentiated it from the rest of the world's was the widespread application of steam engines to pumping water and to operating machinery. John Smeaton was intimately connected with this innovative steam engine technology, and we may be sure that Latrobe absorbed much of his interest. Latrobe's English career also coincided with the fruitful early years of the engine-building firm of Mathew Boulton and James Watt, and he was aware of their work. England was also the world's leader in transportation improvements, particularly canals and turnpikes. Latrobe probably worked under one of Smeaton's pupils as a divisional engineer of the Basingstoke Canal, a waterway near London, and he was a consultant on another canal in Essex. Thus Latrobe was immersed in some of the most recent technical developments of the time.

In architecture, England was the focal point of the Classic Revival. This school emphasized such elements as the dome, porticoes with Ionic and Doric columns, and a simple, unadorned appearance. This style was the major influence on Latrobe's taste, and the private dwellings in England which he designed reflect it.

Latrobe left England for the United States late in 1795 for personal reasons, probably stemming from his wife's recent death, and arrived at Norfolk, Virginia in March 1796. For the rest of his life, he remained in the United States and called himself an American.

Latrobe's best known legacy from his American years is his architecture. While in Virginia he designed his first major work,

35

The Centre Square Engine House of the Philadelphia Water Works exemplifies Latrobe's Classical Revival architectural style (Engraving by W. Birch & Son).

Source: National Archives

the Richmond Penitentiary, a massive masonry structure with a plan reflecting the best penological theory of the day. Moving to Philadelphia in 1798, he had his first opportunities to express his Classic Revival ideas in the Bank of Pennsylvania, generally thought to be one of the most beautiful public buildings in American architecture, and the Centre Square Engine House, an extremely pure and severe example of the style. After Thomas Jefferson appointed him Surveyor of the Public Buildings of the

The U.S. Capitol Building in 1825, painted by Latrobe, who designed its south wing and rebuilt it after it was burned in 1814 by the British.

United States in 1803, Latrobe undertook the redesign and construction of the United States Capitol and the White House, giving them much of the classic beauty which can be seen by visitors today. He also designed many private homes throughout the Middle Atlantic states. In addition to these secular buildings, Latrobe was the architect of several churches, of which the largest and the most impressive is the Baltimore Cathedral. The most exquisite is probably St. John's in Washington ("The Church of the Presidents"), a characteristically simple expression of his taste.

As if such a prolific architectural career were not enough for one man, Latrobe was also the most active engineer in America during the first two decades of the 19th century. Latrobe's greatest engineering commission was the Philadelphia Waterworks, which provided America's largest city at the time with the first comprehensive water system in the United States. He also designed

37

and built the first waterworks at New Orleans. In transportation Latrobe played a significant role. He directed the navigational improvement of the lower Susquehanna River for the state of Pennsylvania, was the engineer of the Chesapeake and Delaware Canal during the first attempt at its construction, and was also engineer of the Washington, D.C., Canal. He was a consultant on a number of canal projects and turned down positions with others. Latrobe also designed or directed industrial establishments, including America's first steam-powered rolling mill, a forge, and a saw and blockmill at the Washington Navy Yard, one of the first steamboat yards in Pittsburgh, and a woolen mill in Steubenville, Ohio.

This staggering list of Latrobe's activities shows that he brought a wide range of knowledge and skills to his adopted nation, and further consideration of his projects indicates that there were several ways in which Latrobe transferred his technical knowledge to America. Perhaps the most striking was Latrobe's transfer of technical concepts by making them believable and acceptable to Americans and persuading people to adopt new ideas. His role in the establishment of the Philadelphia Waterworks is a good example of this process.

Although during the 1790's the city of Philadelphia had been troubled by its dependence for water on contaminated and inadequate wells, it had found no plan for remedying the problem which met with the approval of the citizens or city councils. Mechanical waterworks existed in the United States, but nowhere was a system established which was suitable for a major city. On the other hand, urban waterworks were common in Europe, and Latrobe was familiar with them. While he lived in London he observed particularly the several innovative steam-powered waterworks which supplied the city with river water distributed through wooden pipes under the streets.

Late in 1798 Latrobe came to Philadelphia to erect the Bank of Pennsylvania and attracted the attention of a committee of the city councils which had to look into the water problem. He was asked by the chairman to examine springs north of the city and to determine their suitability for a water supply. The result of his investigation was a report which exceeded his mandate and presented his ideas for a comprehensive waterworks based on English precedents. Latrobe's proposal revolutionized Philadelphia's thinking about how to obtain a better water supply. Although he did consider the springs as a source, Latrobe rejected them as insufficient for the purpose of washing the streets and cooling the air, which he and many residents considered the most pressing needs of the city. Latrobe proposed taking water from the adjacent Schuylkill River, pumping it up by two steam engines to an elevated reservoir and distributing it throughout the city in wooden pipes. He thought that such a system could be ready in seven months.

The most daring element of the plan was the use of steam engines. There were at this time only three steam engines in the United States (none of them in Philadelphia) and no regular steam engine foundry. American knowledge of the capabilities and liabilities of engines was fragmentary. Latrobe attempted to educate the public by reporting all that he could remember of steam-powered waterworks in London and by calling on citizens who had visited England to verify his report. He pointedly described the Chelsea Waterworks, one of the earliest waterworks with one of the famous Boulton and Watt engines. To allay fear and scepticism of steam engines he wrote:

> ...soon after the[ir] invention, steam engines were justly considered as dangerous, man had not yet learned to control the immense power of steam, and now and then they did a little mischief. A steam engine is, at present, as tame and innocent as a clock.

> I have no doubt that this city can produce Smiths capable of constructing very efficient Engines, under proper direction.

The committee of the city councils was taken with the authority of Latrobe's proposal and immediately employed him as consulting engineer. By March 1799, Latrobe contracted for two engines from the Soho Foundry of northern New Jersey, where a former Boulton and Watt engineer was employed. On March 2 the city authorized construction of the waterworks on the basis of Latrobe's estimate of $150,000 for the project.

The critical feature in the city's adoption of this project was Latrobe's informed and enthusiastic presentation of the innovative waterworks concept. His English engineering skills and knowledge permitted him to convince the citizens to solve their water problem by adopting the "latest" technology. In the end,

Latrobe's design proved more costly than estimated and was always a drain on the city treasury. In 15 years it was replaced by a new system, which in seven years was replaced by a third, but in spite of that, Philadelphia remained committed to the concept of a comprehensive urban waterworks and, indeed, was the center of developments in waterworks technology for the first half of the 19th century.

Latrobe also brought to America the professional engineering skills of planning, organizing, and directing technical projects, such as canals. Americans had projected many canals since their Revolution, but few had been started by the time Latrobe arrived in the United States. One well-trained English engineer, William Weston, had preceded Latrobe and contributed his skills to canal projects in the middle and northern states, but he only partially trained a few men to continue his work before he returned to England in 1799. Like Weston, Latrobe found Americans interested in, and appreciative of, his skills in canal engineering. For example, the Washington Canal Company was first organized in 1802, and the directors hired Latrobe to conduct the preliminary survey in the next year. Since he divided his time between public buildings in Washington and preliminary surveys for the Chesapeake and Delaware Canal, he did not write his survey report until early in 1804. When the stock subscription books were opened in June, his report was exhibited as proof of the canal's feasibility. Unfortunately, the canal's promoters were unable to gather enough capital to support the venture.

In 1809 the company was rechartered, and this time a sufficient number of shares were subscribed. The directors again approached Latrobe for advice, but he refused to help unless regularly appointed the canal's engineer. The directors agreed to the arrangement, provided they could deduct from his salary the $300 he had received from his 1804 report. Latrobe acquiesced to the stipulation and also agreed to their condition that he make the canal "on the most oeconomical scale, and in the most oeconomical manner which is consistent with its utility and the provisions of the law." That meant making the locks and other structures of wood instead of stone, Latrobe's favorite construction material. Latrobe in this regard was like virtually all engineers with English training, believing that to build well was to build permanently, and to build permanently required stone. He never forgave the

Washington Canal directors for forcing him to build wooden locks, and he was bitterly pleased when the locks deteriorated within two or three years and the company asked him to rebuild one of them as a stone lock.

Latrobe's belief in English professional standards brought him criticism in other circumstances, but in this case he felt vindicated. He wrote:

> In our haste to be the most enlightened nation in the world, we forget that our very existence is of recent importation from Europe . . . . One part of this error branches out into an opinion that Europeans are expensive merely for the pleasure of being so, that it is a sort of disease incident to their climate as oeconomy is a virtue indigenous in our soils. Oeconomy! The oeconomy of Pendulum mills, & building up the Capital by piecemeal, and canals receiving all the Wash of the land, & wooden locks, & a thousand other cheap things. I am ashamed and angry at it, as an American.

In spite of criticisms of his English engineering attitudes, Latrobe was highly valued as a planner and organizer, and he was employed on many technical projects.

Latrobe not only transferred technology in his own person but also aided the transfer of technology by finding jobs for many skilled Europeans and Americans trained by Europeans. Persons competent to supervise works under his direction were particularly valuable to Latrobe because he often had two or more projects under way at once. Thus for the Philadelphia Waterworks he insisted on the right to select a man paid by the city to supervise daily activities, and he chose John Davis, who had architectural and engineering experience in England. John Lenthall, his man at the United States Capitol, also had an English professional background. For his assistant on the Chesapeake and Delaware Canal he selected Robert Brooke, a Philadelphia surveyor who had worked under William Weston. Latrobe exhibited the same interest in the English talents of the carpenters, masons, and contractors who actually performed the labor. At the rolling mill he hired John Parkins, an iron worker from Sheffield; at the Philadelphia Waterworks he employed Thomas Vickers, a mason who worked under Weston; and on the Chesapeake and Delaware Canal he made a contract with Charles Randall, whom he described as "a practical road and Canal-maker from England."

Latrobe gave many of these skilled people their first American employment, and several went on to long and distinguished careers here. John Davis, for example, left the Philadelphia Waterworks in 1805 to become the engineer of the first waterworks of Baltimore. Others worked for Latrobe on more than one project or were referred by him to other works. Such a man was Charles Randall, whom Latrobe also employed on the Washington Canal, and for whom he obtained a contract on the National Road, the federal highway from Cumberland, Maryland to Wheeling (now West Virginia).

Latrobe also promoted the transfer of skills by training a few pupils in his office. These were young men whom he took in as apprentices, educated in such basic skills as drafting and surveying, and gave valuable experience in supervising his projects. The most important pupils were Frederick Graff, William Strickland, Robert Mills, and Latrobe's own son, Henry Latrobe. By this personal system of tutelage, which was a common means of training engineers in England, Latrobe transferred his knowledge, skills, and opinions to a new generation of engineers and ensured his lasting influence on American technology. Frederick Graff, for example, became the most celebrated waterworks engineer in the country during his tenure as superintendent of the Philadelphia Waterworks from 1805 to 1847. In those years Graff was consulted on virtually every major water supply project in the nation.

Finally, Latrobe contributed to the transfer of technology by his participation in a community of American technicians who were European immigrants or who had visited Europe. Such men as engineer Robert Fulton, politician and architect Thomas Jefferson, chemist and entrepreneur Eric Bollmann, and merchant and papermaker Joshua Gilpin were Latrobe's close friends and frequent correspondents. Their letters to each other are filled with lively discussions of technical innovations.

These friendships led to formal recognition of Latrobe as one of the United States' foremost professional technicians when he was elected to the American Philosophical Society, the first U.S. "scientific" society, and also to the Chemical Society of Philadelphia. Latrobe contributed several papers to the *Proceeedings* of the Philosophical Society, including the first critical essay on American steam engines.

This is not intended to suggest that Benjamin Henry Latrobe was the fountainhead of early American technology, nor even that he must have a prominent place in the pantheon of great American engineers. But there are a number of insights into American technological development which are revealed by the study of Latrobe's career.

Latrobe provides a good example of how new or "hot" technology is transferred. It is curious that the economies of several technically less developed nations have developed rapidly apparently because they adopted the *latest* technology available from the developed nations, rather than simply adopting the standard, tried-and-true technology. This may be accounted for by taking into consideration the enthusiasms which new technology evokes—especially in the trained technician, but also in people generally—leading to the adoption of innovations even when they are at first costly and unprofitable.

Thus the Philadelphia Waterworks required the city government to borrow money and increase taxes during the early years of the waterworks' operation and caused an operating deficit of over a half million dollars in its first 14 years. Yet this patronage brought to Philadelphia a group of mechanics and steam engineers who made Philadelphia the first steam capital of the United States and helped Philadelphia begin its 19th-century career as one of America's centers of mechanical engineering. Perhaps we should reconsider the notion that it is primarily the expectation of profit which originates and sustains technical innovation.

Another important lesson which we may derive from Latrobe's career is the significance of other trained individuals to the transfer of his technical skills. Latrobe was a highly skilled man, bringing with him new concepts and abilities, but he needed other skilled men to understand and execute them. As gigantic as his achievements may seem, they could not have been accomplished without such close assistants as John Lenthall and John Davis, such brilliant pupils as William Strickland and Robert Mills, or such able craftsmen and contractors as Thomas Vickers and Charles Randall. Latrobe's surviving papers reveal dozens of men who were his collaborators, whom he trusted to execute his designs and instructions. Latrobe cannot stand alone as a hero, because his work is the product of many others' active and creative contributions—including many whose names are lost to us. Could we examine all of our technologically great men as closely, we would find how much their accomplishments are inseparable from those associated with them.

The American career of Benjamin Henry Latrobe is a very revealing slice of the early years of the nation. It shows the dependence of the United States on Europe, particularly England, for technical innovation and the way innovations were transferred to America. We find that innovations were enthusiastically received and their introduction involved a large number of persons. We have some notion of how new skills and ideas were transmitted to the next generation. A close look at Benjamin Henry Latrobe's career leaves us more appreciative of the origins of American technical skills.

5

ELI WHITNEY AND THE AMERICAN SYSTEM OF MANUFACTURING*

Merritt Roe Smith

Commonly acknowledged as the "father of American technology," Eli Whitney (1765–1825) towers above other early 19th-century inventors as a technological folk hero. Best known is his invention of the cotton gin, a device which revolutionized agriculture and made cotton king in the plantation South. Of equal technical significance is his association with developing methods of interchangeable production, a seminal part of the Industrial Revolution, which became known during the 1850's as the "American System" of manufacturing. These contributions made Whitney a legend in his time and one of the great men of the modern era.

One thing Whitney never lacked was good publicity. Biographies, popular histories, and textbooks abound with glittering accounts of how he almost singlehandedly inaugurated a "dual revolution" in agriculture and industry, a revolution which purportedly made Americans a "people of plenty" and defined the character of their national life. To generations of schoolchildren Whitney's name became synonymous with the new machine tech-

*Parts of this essay are adopted from my book, *Harpers Ferry Armory and the New Technology: The Challenge of Change*. I am grateful to Cornell University Press for permission to publish this information.—*Merritt Roe Smith*

nology, a democratizing force which helped to stabilize America's shaky economy and legitimize the new republic in the eyes of the rest of the world. Yet, as often happens in the building of heroes, publicists created a myth larger than the man. The purpose of this essay is to re-evaluate Whitney's role as inventor of the American System and place him in the larger context of the 19th-century technological development.

* * *

The origins of the legend date back to the spring of 1798 when the debt-ridden Whitney, his cotton gin business on the verge of collapse, approached Secretary of Treasury Oliver Wolcott about manufacturing muskets for the United States. "I am persuaded," he wrote Wolcott, "that machinery moved by water (and) adapted to this business would greatly reduce the labor and facilitate the manufacture of this article." This suggestion, coupled with the promise of rapid delivery, evidently appealed to the government hovering on the brink of war because Wolcott offered Whitney a contract for 10,000 regulation muskets. Signed on June 21, 1798, the pact called for final delivery by September 30, 1800. The government, on its part, promised to pay Whitney $13.40 a musket and make generous monetary advances at various stages of the contract. Whitney considered the latter provision particularly important. "Bankruptcy and ruin were constantly staring me in the face," he confided to a friend. "By this contract I obtained some thousands of Dollars in advance which saved me from ruin."

Whitney's entry into the arms business undoubtedly was an act of desperation; the threat of "bankruptcy and ruin" drove him into making a very rash proposal. The scale of the undertaking was unprecedented. No American armory, not even the large national establishments at Springfield, Massachusetts and Harpers Ferry, Virginia had ever produced 5,000 arms annually. What is more, Whitney knew little or nothing about arms making. He soon learned that manufacturing firearms was a good deal more complex than building cotton gins. Before a single musket could be produced he had to outfit a factory, recruit a labor force, and procure the necessary raw materials. Just how rash his proposal was became evident when he failed to meet his first contract deadline, the delivery of 4,000 muskets, on September 30, 1799.

46

With less than a year remaining on the contract, he still had not equipped his armory. Even so, through some astute maneuvering he managed to convince federal authorities to grant one extension after another until he finally completed the contract in January 1809—more than eight years beyond the original deadline.

Whitney successfully forestalled cancellation of his contract by cultivating an idea that captured the imagination of government officials. In July 1799 he responded to expressions of concern about not fulfilling his obligation by announcing that he was developing a "new principle" of manufacturing that would not only revolutionize the arms industry but also vastly improve the quality of arms delivered to the government. "One of my primary objects," he wrote Wolcott, "is to form the tools so the tools themselves shall fashion the work and give to every part its just proportion—which when once accomplished, will give expedition, uniformity, and exactness to the whole.... In short, the tools which I contemplate are similar to an engraving on a copper plate from which may be taken a great number of impressions perceptibly alike." Here was the first clear intimation that Whitney planned to make arms with uniform, even interchangeable, parts. The copper plate analogy not only made sense; it convinced Wolcott and his staff that Whitney was indeed on the brink of an important technical discovery.

Having announced his plan, Whitney continued to champion the uniformity concept before members of Congress and other government officials. His most famous demonstration took place at the nation's capitol early in January 1801, before an audience that included President John Adams and President-elect Thomas Jefferson. Acting with typical aplomb, he showed how he could fit ten different lock mechanisms to the same musket with the use of a simple screwdriver. The intimation was clear: Whitney was making firearms with uniform parts. Those present were astonished. Even Jefferson, who was familiar with the earlier work of Honoré Blanc in France, marvelled at the display. "Mr. Whitney," he later wrote James Monroe, "has invented moulds and machines for making all the pieces of his locks so exactly equal, that take 100 locks to pieces and mingle their parts and the hundred locks may be put together as well by taking the first pieces which come to hand." "This is of importance," he continued, "because out of 10 locks, e.g. disabled for want of different pieces, 9 good locks

47

may be put together without employing a smith." Jefferson clearly recognized the military significance of uniformity.

The 1801 demonstration made a deep impression on America's power elite. Thereafter Whitney's name became closely identified with the development of interchangeable parts, an association that eliminated further difficulties in getting extensions on his contract and made him one of the most respected men of his era. Suddenly he found himself lionized by public leaders and hailed as the "Artist of his Country." With recognition came respect, not only as an arms maker but also as a man of affairs whose opinion and advice was sought on a broad range of social and economic issues. A measure of the man's growing popularity could be seen at his Mill Rock armory near New Haven, Connecticut, where scores of curious sightseers flocked to see the inventor and his system at work. Diarists and gazetteers also began to enshrine "Whitneyville" as a monument to American technological genius. Typical was Edward Kendall's *Travels Through the Northern Parts of the United States,* a three-volume work published in 1809 that briefly described Whitney's operations and echoed Jefferson's praise. During the next four decades other writers such as Timothy Dwight (1823), Denison Olmstead (1832), and Henry Howe (1844) embellished the story. From this literature emerged the full-blown Whitney myth incorporated into hundreds of 19th and 20th-century textbooks and taught to millions of schoolchildren.

* * *

Whitney's reputation as inventor of the American System stood unchallenged until the 1960's when researchers began to detect serious discrepancies between the written record and extant artifacts. Most damaging was the discovery that Whitney muskets, including those made under his first contract, did not have interchangeable parts. The physical evidence was unmistakable. "In fact," one scholar exclaimed, "in some respects they are not even approximately interchangeable!" This, coupled with the fact that individual lock components of Whitney muskets carry special identifying marks—something which is unnecessary for truly standardized parts—indicates that Whitney must have staged his famous 1801 demonstration with specimens specially prepared for the occasion. In short, it appears that Whitney purposely duped government authorities in 1801 and afterwards encouraged the notion that he had successfully developed a system for producing uniform parts.

Is the Whitney legend completely fabricated? If Whitney did not make arms with interchangeable parts, should he be written off as a charlatan and dismissed as having played no part in the development of the American System? The answer to both questions is emphatically *no*. Clearly part of the legend is salvageable, but an accurate understanding of Whitney and his work requires a reassessment of the man as innovator, advocate, and symbol. Such an analysis not only addresses problems of technical creativity but also sheds light on the complex social and ideological character of 19th-century American industrial development.

Mechanization was a vital part of the American System, and Whitney recognized its importance at an early date. Indeed, his initial emphasis on machine production, alluded to earlier, was instrumental in gaining his first contract in 1798. Yet, curiously enough, a survey of equipment at Whitney's Mill Rock armory reveals a very modest attempt at mechanization. The articles listed are not nearly as sophisticated as one might expect to find in a pioneering "best practice" establishment.

In September 1801, a few weeks before Whitney made his first delivery of 500 muskets under the 1798 contract, young Philos Blake wrote home about the machinery he had seen at his uncle's factory. The letter mentioned a drilling machine, a boring machine, a screw machine, and a trip hammer. All four devices were well known to Whitney's contemporaries. The most interesting unit was the screw machine, a device for hollow-milling screw forgings which Whitney may have copied from the Frenchman Blanc. By 1825, the year of his death, Whitney had added to his mechanical repertoire a stamping machine, a "nitching" machine for slotting screw heads, a pair of cast iron shears, and two polishing machines. In terms of design, none of these machines represented new concepts. Even the famous "Whitney" milling machine, a self-acting device employing toothed cutters capable of producing both plain and irregular surfaces, is now known to have been built not by the master but by his nephew, Eli Whitney Blake, around 1827 and probably copied from a similar "straight cutting engine" designed by John H. Hall of the Harpers Ferry armory.

Mill Rock was a transitional establishment, a halfway point

between the traditional craft shop and the fully mechanized factory. The physical plant and labor force were never large (the latter averaging around 50 workers), and production rarely exceeded 1,500 muskets a year. Compared with other New England armories circa 1820, Mill Rock's machinery was neither plentiful nor particularly distinctive. Whether built on the premises or acquired from others, the designs reflected simplicity and cheapness. In fact, the conspicuous absence of machinery for performing the weightier tasks of gunmaking reveals that Whitney purchased a large number of components from outside sources. From the very beginning he had subcontracted for finished gunstocks and barrels, and he continued the practice into the 1820's. He intentionally confined his operations to the on-site manufacture of lock mechanisms, unquestionably the most complicated part of the musket but also the one part that could be produced without a large investment in heavy machinery. This strategy fully comported with Whitney's view of Mill Rock as an assembly shop rather than a fully integrated manufacturing establishment. He made his muskets almost entirely by hand methods, and his limited stable of machinery accentuated this orientation.

That labor-intensive methods dominated production at Mill Rock is clearly shown by the large number of files carried on the armory's inventory. Indeed, hand filing represented the touchstone of Whitney's manufacturing system. The filing shop accordingly became the focal point of activity. Visitors invariably commented on the "moulds" used by workers at Mill Rock, attributing much of Whitney's success to their widespread adoption. Commonly known as "jigs," these implements consisted of tempered steel patterns which guided armorers in filing down components to their proper dimensions. Their use was not new, even though Whitney claimed to have invented the technique and extended its applications far beyond anything achieved by his contemporaries. Ironically, physical limitations in the jigs themselves prevented him from achieving what he had promised government authorities in 1801—the production of muskets with interchangeable parts. As it was, jig filing allowed Whitney to substitute unskilled for skilled labor in the productive process and thereby effect substantial cost savings. Herein lay the true significance of his manufacturing system.

A distinguishing feature of industrial development is work

specialization through the division of labor. When labor is divided, as in jig filing, individual work assignments become more simplified while the overall production process becomes more complex. The need for coordination and control is therefore more urgent, and this, in turn, intensifies managerial responsibilities. Further demands are placed on managers when the eventual growth of an enterprise increases the speed and scale of production. Whitney, to his credit, recognized this problem and began to establish procedures for monitoring the work process at Mill Rock as early as 1798. His efforts took two forms: cost analysis and labor regimentation. Both aimed at improving the efficiency and flow of production, and both constituted important managerial advances.

A shrewd businessman, Whitney knew that proper attention to costs could mean the difference between profit or loss, success or failure. Unlike other private arms makers who simply estimated costs by adding a dollar or so to the going rates for labor and raw materials, Whitney sought to specify costs with precision. During his 1807 contract negotiations with the War Department, for example, he figured interest as well as insurance charges in calculating the price of his product. By 1812 he was beginning to add depreciation allowances for "wear and tear" on his plant and machinery. Coupled with the use of piece-rate accounting, his method of estimating costs placed him in the vanguard of factory masters who were trying to rationalize production through systematic procedures and regularized practice. As a result, Whitney enjoyed lucrative profits from the firearms business while many of his less advanced contemporaries found themselves operating at the margin. This aspect of Whitney's career has received little scholarly attention, but preliminary research indicates that his contributions to management are quite significant and certainly overshadow the mechanical inventions for which he is best but undeservedly remembered.

Whitney's desire for order and discipline applied especially to the people he employed. Like other early factory masters, he styled himself as a "steward of the Lord" whose special calling was spiritual as well as worldly leadership. Seen in this context, Mill Rock became not only a place of production but also a moral gymnasium where "correct habits" of diligence, thrift, and sobriety were inculcated through the systematic control of everyday life. Such an orientation allowed Whitney to amalgamate the sacred and the profane, thereby identifying godliness with productivity and dealing a serious blow to agrarian critics who railed against the moral decadence of industrialization as a threat to republican virtue.

The Eli Whitney Gun Manufactory at Mill Rock, Connecticut, attracted many visitors to see Whitney's manufacturing and management innovations (Painting by William Giles Munson).

Source: The Mabel Brady Garvan Collection, Yale University Art Gallery

William Giles Munson's painting of the Mill Rock "Gun Manufactory" reflects Whitney's penchant for social control. Everything in the bucolic setting centers around the neat and orderly armory. Even elements in the landscape—the mill dam, the covered bridge, the symmetrical arrangement of worker housing, the fenced pastures, the rich foliage—convey a sense of balance and stability. Life within the village paralleled this orderly view of the universe. The factory bell called armorers to and from work as clocked time gradually supplanted irregular patterns of toil. Rules punished indolence and encouraged diligence while the strict regimen itself systematically eliminated individuality from the productive process. Above all, prohibitions against drinking, loitering, swearing, and other "improprieties" during working hours eroded pre-industrial traditions that had formerly mixed life and labor. Outside the shops armorers and their families were expected to observe the Sabbath and lead moral lives. Those who transgressed established norms were chastized, fined, and in some cases deprived of credit and housing. Repeat offenders were fired and blacklisted, their chances of finding employment elsewhere in the region virtually eliminated. These and other sanctions set a pattern of paternalism that would be repeated time and again in 19th and 20th-century America.

Despite his public charisma, Whitney was not popular with his employees. Armorers chafed under the pressures of working in a closely regimented environment. Constant supervision accompanied by carrot-and-stick methods gave rise to anxieties that vented themselves in labor disturbances and expressions of discontent. Not surprisingly, Mill Rock had one of the highest labor turnover rates in the arms industry. On several occasions Whitney posted bills for the return of runaway apprentices. Far more frequently, dissatisfied armorers simply left their jobs for more appealing positions elsewhere. Those who remained at Mill Rock continually grumbled about low wages and the rigidity with which foremen enforced the rules. Serious morale problems existed at Mill Rock, and Whitney's inability to attract and retain competent workmen severely hampered his efforts to attain the engineering ideal of uniformity.

Even though Whitney sought to develop new manufacturing and management methods, his efforts fell short of success. Except for a promising attempt at cost analysis, he introduced little that was startlingly new at Mill Rock. From filing jigs to hollow mills, the techniques he used originated with others. Yet his zealous advocacy of uniform production made him one of the key progenitors of the American System. From the start he not only popularized the uniformity idea but also persuaded politicians to support policies aimed at standardizing the manufacture of military arms. Time and again he conferred with ordnance and armory officials regarding the preparation of model arms and the introduction of new techniques. A frequent correspondent was Roswell Lee, a former protégé whose innovative management made the Springfield armory the hub of the American firearms industry. Because of his reputation and influence, Whitney participated in virtually every important government decision bearing on the manufacture of firearms between 1801 and 1825. In June 1815, for instance, he hosted a meeting at New Haven, Connecticut that devised a formal strategy for standardizing the manufacture of military muskets at the national armories. Present were Lee, Benjamin Prescott (Lee's predecessor at Springfield), superintendent James Stubblefield of the Harpers Ferry armory, and Colonel Decius Wadsworth, chief of the U.S. Ordnance Department and one of Whitney's closest friends. Out of this meeting came the decision to introduce a new Model 1816 musket as well as special inspection gauges at the national armories and eventually to distribute them to private contractors. The adoption in 1823 of these gauges, though primitive by modern standards, signalled the beginning of a sustained effort toward the production of truly interchangeable parts, a goal that would eventually be achieved during the 1840's.

In the final analysis, Whitney's most important contribution to the American System stemmed not from any technical innovations he made but from his active promotion of the engineering ideal of uniformity. He was, to be sure, an entrepreneur who saw opportunities for profit and took them; but he was also an *advocate* who perceived the larger benefits of precision manufacture and worked assiduously to keep the concept before the public eye. Beyond this, his conscious cultivation of the uniformity system had a larger cultural purpose. At a time when the fledgling United States badly needed psychological anchors and a collective social vision, Whitney helped to identify technical progress with political and economic stability. In his mind technology meant something more than a means of controlling nature and producing goods; it represented a barometer of national development, an accomplishment in which Americans could take pride and from which republican institutions could draw strength and vitality. Through technology, then, everything associated with the American Dream—material abundance, democracy, freedom, perfectability, destiny—could be achieved. In time Whitney came to symbolize this union. Along with Samuel Slater, Oliver Evans, and other lesser known individuals, his name not only stood for material progress through technology but also for republican virtue in an era of developing industrialism.

*   *   *

54

At least three major streams of inventive activity converged during the 1820's and 1830's to produce the mechanical synthesis known as the American System. Two centers of this activity —Middletown, Connecticut and Springfield, Massachusetts— were located along the Connecticut River within 40 miles of each other and shared a common heritage going back to the earliest Puritan settlements of the 17th century. The third center stood near the junction of the Potomac and Shenandoah rivers at Harpers Ferry, Virginia and reflected mores and traditions quite different from those of New England. Yet, despite these cultural contrasts, all three communities developed exciting new technologies, built on one another's advances, and eventually succeeded in manufacturing firearms with interchangeable parts. Together they exercised a cumulative impact on American metalworking and shop practice that prevails to this very day.

Middletown enjoyed a widespread reputation as the home of a number of distinguished arms makers. Among the most prominent were Henry Aston, Robert and J. D. Johnson, Edward Savage, and Nathan Starr and son. Yet, as able as they were, none of these men displayed the mechanical ingenuity and skill of Simeon North (1765–1852).

A quiet, unassuming person, North had a very prolific career as an arms maker. He secured his first government contract in 1799 and continued as a private contractor, manufacturing horsemen's pistols, common rifles, and Hall breechloaders, until his death in 1852. Apart from his remarkable longevity, North possessed a talent for technical creativity that made him a trend setter. At a time when most armorers relied almost exclusively on hand tools to execute their work, he sought to mechanize his operations and adopt precision standards. Success came early. By 1816 he had not only equipped his "Staddle Hill" factory in Middletown with advanced gun-stocking and barrel-turning machinery but had also succeeded in manufacturing pistol locks with unmarked interchangeable parts. Equally significant, he had begun to form some of the small lock components with a plain milling machine—the first of its kind in America—thereby effecting significant cost savings in file consumption as well as higher standards of accuracy. Ordinarily these accomplishments should have catapulted North to fame and fortune. But lacking Whitney's flamboyance and charm, he failed to attract public notice. While recognized by peers as an ingenious inventor, he died as he had lived—in relative obscurity.

What North initiated at Middletown, John H. Hall extended at Harpers Ferry. A native of Portland, Maine, Hall (1781–1841) received a government contract and moved to Virginia in 1819 to manufacture a breechloading rifle bearing his name. At the outset he asserted that he would make his rifles with interchangeable parts, a point scoffed at by skeptics but strongly supported by ordnance officials who emphasized the experimental nature of the work. Provided with makeshift facilities at the Harpers Ferry armory, between 1821 and 1827 Hall built an impressive stable of machinery that included water-powered drop forges as well as several types of milling, drilling, and profiling machines. Featuring automatic stop mechanisms, formed cutters, and total metal construction, this equipment represented a significant step forward in mechanical design. Attention to small details, such as the use of balanced pulleys, not only increased the speed and efficiency of operation but also extended engine life as well. Beyond this, Hall sought to insure precision by executing work from special bearing points, adopting standard screw threads, and using over 63 gauges during the manufacturing process. These innovations enabled him to achieve tolerances of one-thousandth of an inch and in 1826 successfully complete the first fully interchangeable weapons ever made in the United States.

Although Hall conducted his operations in a small pilot plant and failed to achieve significant economies of scale, he is nonetheless a pivotal figure in the annals of industry. What was so startlingly new about his work was the extent to which he mechanized his operations and the impressive results he actually achieved. No one at the time Hall went to Harpers Ferry had been able to master the problem of attaining complete interchangeability in firearms. Much of the excitement generated by his work can be traced directly to his success in combining men, machines, and precision-measurement methods into a practical system of production. In 1819 no one—not even the chief of ordnance—was absolutely certain that mechanized processes could be successfully applied to every aspect of gun making. Hall showed that they could and, in doing so, bolstered confidence among arms makers that one day they would achieve in a larger, more efficient manner what he had done on a limited scale. In this sense, Hall's work represented an important extension of the Industrial Revolution in America, a mechanical synthesis so different in degree as to constitute a difference in kind.

Like Hall and North, armorers at the Springfield armory played an indispensable role in fashioning the American System of manufacturing. Under the dynamic and resourceful leadership of Roswell Lee (1815–1833), Springfield not only became the "Grand National Armory" but also emerged as one of the most progressive manufacturing establishments in the United States. Lee recognized that, no matter how refined manual skills became, muskets made by hand simply could not compare in accuracy and workmanship with those finished by machinery. Process refinement consequently became the keystone of his policy, and from his appointment as superintendent in 1815 to his death in 1833, he doggedly pursued the mechanized production of uniform parts. Although complete success eluded Lee during his lifetime, he nonetheless established a managerial strategy at Springfield which culminated during the 1840's with the design and manufacture of the U.S. Model 1842 musket. This was the first regular-issue weapon ever to be mass-produced with interchangeable parts and, as such, constitutes a milestone in the history of technology.

Springfield's leadership in the mechanical arts depended less on in-house inventive activity than on an uncanny ability to monitor and assimilate work being done elsewhere. Lee and his successors, John Robb and Major James W. Ripley, excelled at ferreting out technical novelties and incorporating them into the productive process at Springfield. Because the most important private arms makers—like Simeon North and Asa Waters—depended almost exclusively on the federal government for contracts and because ordnance officials insisted that contractors share new technological information with the national armories, Lee and his staff had access to virtually every important workshop in New England. Inspectors and machinists from Springfield were continually being dispatched to private armories during the 1820's and 1830's to examine new machinery and make drawings so that the devices could be replicated at Springfield. Through this intelligence network Lee procured invaluable information on the latest techniques of milling, forging, and gauging. Through the same network he learned about Thomas Blanchard, a young mechanic from Millbury, Massachusetts who in 1818 devised an ingenious cam-operated lathe for turning the irregular surface of gun barrels. After conducting a demonstration at Springfield, Blanchard not only installed the machine at the national armories but went on to apply the principle of eccentric turning—tracing a master pattern and reproducing it—to the manufacture of gunstocks. Lee deemed this invention so important that he subsidized Blanchard by bringing him to Springfield in 1822 as an inside contractor. During the next five years the Millbury mechanic designed and perfected 14 different woodworking machines which completely mechanized the process of gunstocking, insured uniformity, and eliminated the need for skilled labor in one of the three major divisions of armory production. Rarely have the contributions of one person effected such a sweeping change in manufacturing practice in so short a time.

The presence of Blanchard and other gifted mechanics at Springfield helps to explain the armory's surge to international prominence during the pre-Civil War period. As early as 1825 Springfield had become the linchpin of the American arms industry and a pivotal clearinghouse for the acquisition and dissemination of technical know-how. If an arms maker, or any other manufacturer for that matter, wished to learn about the latest developments in metalworking, he invariably visited the national armory. There he could examine the most advanced tools and machinery, compare notes, and discuss common problems with such master machinists as Adonijah Foot, Cyrus Buckland, and Thomas Warner. After satisfying his curiosity and determining his needs, he might then set out for other establishments to gain further knowledge about new and interesting mechanical ideas. The ability of arms makers to move freely from one shop to another contrasted with more closely guarded practices in Europe. Such an "open-door" policy doubtlessly helps to explain the highly integrated character of the American arms industry as well as the relative speed with which manufacturers assimilated the new machine technology. The same policy also accentuates the public-service orientation of the U.S. military service and armory system during the antebellum period.

By the 1840's, manufacturing techniques aggregated and synthesized at Springfield began to spread to other armories and machine shops throughout the United States. Frequently this influence manifested itself through workmen who had received their early training at Springfield and subsequently moved to new positions as machinists and supervisors at other manufacturing estab-lishments. These "graduates" of the armory became key dis-seminators of the new technology and could be found in most of the important metalworking centers in America. In 1842, for instance, Thomas Warner, one of the architects of the system at Springfield, resigned his position as master armorer and moved to the Whitney armory at New Haven. Several years later the Ames Manufactur-ing Company lured a skilled machinist named Jacob Corey Mac-Farland to its factory at Chicopee, Massachusetts. Like Warner, MacFarland possessed intimate knowledge of the techniques used at Springfield and as foreman of the Ames machine shop intro-duced "armory practice" at Chicopee. Other manufacturers fol-lowed similar strategies, thereby acquiring the latest armory know-how and relieving themselves of lengthy time lags and costly expenditures associated with learning-by-doing. Such methods of transfer and convergence carried mechanized production to a heightened level of maturity during the 1840's and 1850's, paving the way for the rise of the American machine tool industry and its entry into European machine markets prior to the Civil War.

The experience of the Ames Manufacturing Company aptly illustrates how private firms availed themselves of the new tech-nology. Established in 1834 by the brothers Nathan P. and James T. Ames, the company became one of the first businesses in the United States to manufacture and market a standard line of machine tools. The firm also undertook special jobbing contracts and made a wide variety of millwork, mining equipment, cutlery, small arms, cannon, statuary, and other metal castings. Yet, de-spite their obvious technical skill and versatility, the Ames brothers were basically copyists rather than innovators. And for good reason. Chicopee was only a few miles from Springfield and the brothers had ready access to patterns and drawings owned by the national armory. Aided by a number of former Springfield armorers, their stock of machinery clearly reflected this influence.

Through the marketing activities of Ames and other early machine tool companies, armory practice began to spread to tech-nically related industries and by the 1850's could be found in factories making sewing machines, pocket watches, padlocks, and hand tools. From these modest beginnings it was only a matter of time before the new technology would be adapted to the produc-tion of typewriters, agricultural implements, bicycles, cameras, automobiles, and a whole host of products associated with the mass-production industries of the 20th century. Interestingly enough, mechanized techniques rather than precision processes constituted the most important transfer from the arms industry. Since precision production was expensive, most businessmen contented themselves with manufacturing highly uniform but not necessarily interchangeable parts. Only the government could afford the luxury of complete interchangeability.

The capstone of two generations of creativity in the American firearms industry came in 1851 at the London Crystal Palace Exhibition. For the first time little-known Yankee manufacturers impressed visitors with the quality and precision of their machine-made products. Exciting considerable comment and admiration were the unpickable padlocks of Alfred C. Hobbs of New York, Samuel Colt's revolving pistols, and six completely interchangeable rifles made by Robbins & Lawrence, a small "armory practice" company from Windsor, Vermont. All three firms won medals at the exhibition, but more significantly the excellence of their wares prompted the British government to send an investigatory commission, headed by Joseph Whitworth and George Wallis, to New York's Crystal Palace Exhibition and, from there, on a fact-finding tour of the northeastern section of the country during the summer of 1853. The following spring Her Majesty's Ordnance Board, eager to take advantage of the American System, dispatched another three-man "Committee on Machinery" to the United States for the express purpose of introducing similar improvements at the Enfield armory near London. By the time the investigators returned home in August 1854, they had placed orders for over $105,000 worth of machinery, chiefly with Robbins & Lawrence and the Ames Company. Within four years several other governments sent similar investigatory teams to the United States. Although little noticed at the time, these events signaled America's coming of age as an industrial power.

Eli Whitney doubtlessly would have been pleased with the outcome. Although he failed to achieve interchangeable standards during his lifetime, he could take considerable pride in the fact that he had been instrumental in popularizing the engineering ideal and fixing the concept in public consciousness. Through Whitney the uniformity system became a cornerstone of Ordnance Department policy. Through the Ordnance Department, creative mechanics like Hall and North received government contracts enabling them to pursue the elusive goal of interchangeability. Through Hall, North, and other contractors the new techniques coalesced at the Springfield armory; from Springfield they filtered through the nascent machine tool industry to permeate American and European markets. Out of this complex tapestry of invention, cooperation, and diffusion can be discerned a mechanical genealogy that directly links Whitney and other early arms makers with the mass production industries of the 20th century. The achievement sealed the fate of pre-industrial culture in America.

6

THOMAS P. JONES AND
THE EVOLUTION OF
TECHNICAL EDUCATION

Bruce Sinclair

There is a long tradition of popular science education in the United States, and it reflects two important tenets of the national faith. One is that technology—science applied to practical purposes —is central to the nation's destiny; the other that education is crucial to democracy. Few individuals were more clearly identified with these doctrines than Thomas P. Jones, spokesman for a movement of educational reform, editor of the *Journal of the Franklin Institute* from 1826 until 1848, and longtime commentator to the nation on American inventions. In his position as editor of the country's most influential mechanics' magazine, Jones championed the cause of mechanics' institutes and lyceums, apprentices' libraries, and all those similar associations that sprang up in the Jacksonian era to provide useful information to the working classes, so that they might rise to "their proper rank in a republican society."

Although born in England and educated there as a physician, Jones's varied experiences proved remarkably appropriate to a career promoting the development of American ingenuity. His sentiments for reform were undoubtedly stimulated by an association with Joseph Priestley, the celebrated English scientist, political radical, and religious dissenter, with whom he may have emigrated to the United States around 1794, and perhaps that connection also turned Jones's interests to science. During the next 20 years he accumulated a considerable reputation for his lectures on various aspects of science and its applications. His talent for lecturing attracted large audiences and gave him an appreciation of the popular demand for scientific knowledge and some positive ideas about the ways it should be conveyed. In the period before he became editor of the Franklin Institute's magazine, Jones also gained academic experience teaching the principles of science—he was Professor of Natural Philosophy at the College of William and Mary for four years—and he learned the practical skills of the mechanic arts from a partnership in a Philadelphia fire engine manufacturing firm. These activities made Jones well known within the city's community of skilled craftsmen, and when the managers of the Franklin Institute wanted to begin a publication, he was the only person they thought of to edit it. They encouraged him to buy the *American Mechanics' Magazine* from its New York publisher and named him to the professorship of mechanics in the Institute as a further inducement to locate the magazine in Philadelphia.

Jones designed the periodical, which he named the *Franklin Journal and American Mechanics' Magazine,* to be broad in its appeal and purposes. He had two great ambitions for his magazine: that it give workingmen plain and practical information and that it direct American inventive activity into constructive channels. This latter dream took on additional potentiality when Jones was appointed head of the U.S. Patent Office in 1828. The connection seemed ideal. Many Americans thought of the Patent Office as a sort of educational agency, and his friends considered Jones perfectly suited to it, both for his learning and his practical outlook. Moreover, the Franklin Institute had just assumed ownership of Jones' periodical, and his new official position promised national significance for the *Journal.* Indeed Jones himself saw the magazine as the way "to lay open those stores of the genius and skill of our countrymen which, although existing in the Patent Office, have hitherto been but very partially known." In other words, the Franklin Institute's journal would also become the Patent Office's journal. To mark the importance of the relationship, the Institute appointed Jones editor for life, started a new series of the magazine, and changed its name to the *Journal of the Franklin Institute of the State of Pennsylvania, Devoted to the Mechanic Arts, Manufactures, General Science, and the Recording of American and Other Patented Inventions.*

The editor's official connection to the Patent Office proved to be surprisingly brief, however. After only a year he was forced out in

The Franklin Institute in Philadelphia was founded in 1824 for the promotion of the "mechanic arts" (Steel engraving circa 1830 by Fenner, Sears & Company from a drawing by C. Burton).

Source: Library of Congress

one of Andrew Jackson's administrative shake-ups. But Jones remained a familiar figure around the Patent Office for another two decades. He conducted a patent agency for a time, was a patent examiner for a short period after the Patent Office was reorganized in 1836, and throughout the years from 1828 to 1848 he brought all his mechanical and journalistic skills to the analysis of new patents for the *Journal of the Franklin Institute*.

Jones conducted the magazine with simple editorial rules. He believed that *Journal* articles should be couched in "a style as familiar, and as little technical, as the subject will admit." Plain language was what practical men understood best, he argued; scientific or technical terms quite familiar to an author would still be "dog latin to 99/100 of his readers." Jones was also convinced that variety was essential and that articles must be short. Anything longer than eight pages, he claimed, would be ignored by two-thirds of the magazine's subscribers. He had firm ideas about the appropriate contents for his *Journal*, too. Essentially a teacher, his principal aim was to spread knowledge, not to create it. He excerpted material from foreign technical journals, reported on American industrial advances, and encouraged workingmen to share their own insights—but always from the point of view that practical men needed practical information.

Dr. Jones was concerned with something different than an effort simply to raise the level of traditional craft practices, however. He was convinced that education was crucial to the social and economic improvement of artisans, "a class of our fellow citizens," he once wrote, "whose importance we are only beginning to recognize." And he also believed that scientific principles lay at the basis of the mechanic arts. If working people would learn those principles, if they could unite natural laws and their practical skills, they could become "scientific" in their work. This new approach to technical education also shared the values of science. It depended on literacy and an open, cooperative style of work, unlike the secrecy that had dominated craft activity for so long. The utility of such a program seemed plain to Jones, and he dedicated his *Journal* to it. Working people ignorant of first principles wasted time and money on things that would not work. Worse yet, they were misled by mechanical chimeras. As he said in an article on perpetual motion, "there is scarcely any other subject so familiarly spoken of, and so little understood, as the principles of mechanics, and no one, therefore, in which quackery is more certain of success."

Dr. Jones brought the same convictions and insights to his analysis of new patents. He envisioned the Patent Office as a great repository of technical wisdom. He saw it, on one hand, as a museum in which the mechanic could trace the historical progress of the art and, on the other hand, as a collection which described the present state of that art. If properly studied, the materials in the Patent Office would not only save the worker from repeating the mistakes of his predecessors, he would be led to perfect existing processes and to devise new procedures. If the mechanic arts could command their history, Jones believed they could also command their future.

Much of the attention Jones gave to patents in the *Journal's* columns was to diffuse the knowledge they contained. The publication of patents was by itself a useful service, since the Patent Office did not begin a journal of its own until 1843. But in his self-appointed role as public commentator on new inventions, Jones provided a range of additional information to the *Journal's* readers. The specifications of patents he thought particularly interesting were published in full, sometimes with explanatory diagrams and often with further editorial observations. For example, in his list of American patents granted in September 1828, Jones printed the complete specification of a patent for an improved seed planter, including a plate showing the machine and its component parts. In his "Remarks" Jones included the report of a correspondent who had used the planter and his own opinion that the device, developed in New York, would prove especially useful in cotton planting, providing the seed release mechanism were modified along lines he then suggested. Thus the mechanic community had called to its attention a mechanically novel device that had been field tested and that might be altered to serve various purposes.

Teaching was one of Jones's natural inclinations, and he singled out cases such as the seed planter to illustrate useful and progressive inventive activity. More often, however, he was forced to scold inventors who failed to study the lessons of the past or who directed their efforts into irrelevant channels. Of a patent for a revolving belt saw, the editor concluded:

> It is mentioned in Rees' Cyclopaedia; has previously been patented more than once in the United States; has been repeatedly tried, and as frequently abandoned as worthless in

operation; and such will again be its fate, should it again be essayed by the present patentee.

Patents for steam bed-bug destroying machines, for instruments to cure dyspepsia, and, in seemingly endless numbers, for improved washing machines and butter churns, also stimulated Dr. Jones to comment. "This is a swinging, or pendulum churn," he said of a combined churn and washing machine, "which really has some novelty about it, and even in this fact alone, there is something cheering, as originality has long been a rare element in churns and washing machines."

What most exasperated Jones were the mechanical frauds, whether or not conceived in ignorance. "This is a most absurd project," he said of Obed Marston's machine to propel mills, "at least as absurd as the attempt at perpetual motion itself." The specification was also defective, and the patentee had sworn both the oath for citizens and that for aliens. "But even this double fortifying," Jones wrote, "will be of no avail in the attempt to turn dross into gold." Jones used these occasions to stress the importance of scientific principles, natural laws which no jargon could overcome. But he was aware that working people were not the only ones deluded in such schemes. By a special act of Congress, Horatio Spafford received a patent in 1832 for "Discoveries in Natural Philosophy, reduced to practice." Spafford's device contradicted the most elementary laws of nature, Jones said, disgusted "that any person having any pretensions to mechanical knowledge, should be, for a moment, deceived by such a plan of obtaining power." But in fact, Jones knew that some prominent Philadelphians had invested in the project, and he therefore determined to act in such cases "as a public sentinel."

The *Journal's* monthly list of American patents, "With Remarks and Exemplifications by the Editor," became one of its most popular features. Enlivened by Dr. Jones's puckish language and authoritative posture, the column was widely read and extensively quoted. Americans had great expectations for the Patent Office. Mark Twain caught that spirit in *A Connecticut Yankee in King Arthur's Court* when he had Hank Morgan say:

> The very first official thing I did, in my administration—and it was on the very first day of it, too—was to start a patent

office; for I knew that a country without a patent office and good patent laws was just a crab, and couldn't travel any way but sideways or backwards.

Americans imagined themselves an inventive people; indeed, they saw ingenuity as a national characteristic, and they expected the Patent Office to be a monument to that talent. In the self-conscious way they often thought about such things, the patent system would be evidence of their prospects, just as Europe's ancient ruins were testimony to its past. The framers of the Constitution had, from the outset, associated patents with the promotion of "the progress of Science and useful Arts," and thereafter Americans commonly linked invention with the nation's advancement.

Because they used technology to measure their country's development, Americans also made folk heroes of their inventors and elevated technical education to a place of esteem. It became the ideal form of democratic instruction, especially for those who had historically been denied much education from circumstances of birth or occupation. In a sense, useful knowledge became patriotic knowledge, and Dr. Jones could argue with sincerity that his magazine helped sustain "an important part of the fabric of our independence."

These cultural values were reinforced by increasing demands from the marketplace for new kinds of technical education. Craft traditions had never provided a successful form of training. Benjamin Franklin's early escape from apprenticeship was typical of the system's failure in America. Instead, a variety of educational devices emerged in the 18th century: voluntary self-improvement groups, of which Franklin's Junto was simply the most famous; evening schools that taught applied arithmetic and other practical subjects to urban working people; and at an even more popular level, applied science lectures of the sort Dr. Jones had so successfully purveyed. By the early 19th century, however, it was becoming clear that this miscellany of technical instruction no longer suited the nation's requirements. Large-scale civil engineering projects to construct canals and railroads and complex industrial development that depended on steam power and iron machines called for more sophisticated and systematic engineering education.

Dr. Jones was an important figure in this transitional period of technical education in the United States. He perceived two central flaws in the existing educational system. The first was that elementary school instruction was not sufficient to prevent serious differences in educational levels between those who could afford private schooling and those who could not. The second was that apprenticeship systems failed to encourage the intellectual flexibility upon which technical progress depended. He therefore argued for an approach that linked practical and theoretical instruction to a vision of social mobility. For Jones, and indeed for the mechanics' institute movement generally, the solution was to develop new forms of education—low enough in cost to be within the reach of anyone, but without the stigma of public charity schools—more advanced than the elementary schools, but with an emphasis on science rather than the classics. To achieve these goals, most mechanics' institutes offered inexpensive evening lectures, organized into courses of instruction in the principles of physics and chemistry and their applications to the mechanic arts. Mechanics' magazines were infused with the same ambitions, and in his journal Dr. Jones experimented with a variety of special features to teach working people a stock of simple but essential scientific principles. And because he assumed that artisans educated in the principles of science would transcend craft boundaries, he also concluded they would transcend the social and economic limits that had historically circumscribed their lives.

In fact, the celebrated union of theory and practice, for which Jones had held such hopes, turned out to have little more than rhetorical value. The principles of science proved unrelated to the concerns of most artisans. They wanted free public education, mechanics' lien laws, and the abolition of debtor's prisons more than the fundamental laws of physics—which in any event had little evident applicability at the level of daily workshop practice. But Dr. Jones and his fellow reformers were correct to perceive the need for theory in technical education and that new institutional forms were required. They were also correct, in an era before the wide establishment of tax-supported public schools and at a time when private colleges and universities were still dominated by classical traditions, to connect technical education to democratic ideology.

The form that American technical education took embodied

both those perceptions. Within a decade after Dr. Jones's death, the mechanics' institute movement had been translated into a crusade for university-level training in technology for the sons and daughters of ordinary farmers and mechanics. Also known as the "People's College" movement, this interest in publicly supported universities of practical science culminated in the Morrill Act of 1862. The act provided the basis for a nationwide system of publicly supported colleges devoted to scientific instruction in agriculture and the mechanic arts—so-called "land-grant" institutions because the Congress granted federal lands to the states to maintain the colleges. Indeed, in their mixture of practical purposes, applied science, and democratic aspirations, the land-grant colleges gave substance to many of the dreams Dr. Jones had for education in his adopted land.

7

CYRUS HALL McCORMICK AND THE MECHANIZATION OF AGRICULTURE

Carroll W. Pursell, Jr.

American agriculture at the beginning of the 19th century was largely a hand craft. Only a few tasks, such as plowing and hauling, were widely aided by the power of horse and ox. By the end of the century many more tasks—most notably those connected with harvesting grass and grains—were done by machines powered usually by horses or mules but sometimes by steam traction engines or even electricity. The power bottleneck remained—to be broken only after 1900 by the gasoline-powered tractor—but the machines to mow, rake, and thrash were already at work. An industrial revolution was begun in agriculture, and no small part of the credit was due to the inventor and manufacturer, Cyrus Hall McCormick.

As one of the oldest and most fundamental technologies, agriculture had undergone several revolutions over the centuries. The first European settlers in America brought with them a powerful set of devices and techniques which had appeared during the Middle Ages—most notably the horse collar, the heavy plow, and the three-field system of crop rotation. In the new American environment they found soil and climate not too different from those of their mother countries, but at the same time they met a set of circumstances that militated against the simple and direct transfer of Old World technology and techniques to the new. For one thing, most of the accessible land was covered with a dense forest which made the plowing of neat furrows through tidy fields an almost hopeless task. For another, the urgent demand for foodstuffs and export cash crops, coupled with the vast abundance of land as yet uncultivated, made crop-rotation schemes, especially those which allowed some fields to remain fallow for periods of time, seem extravagant and unnecessary.

In the short run, the native American Indians taught the new settlers how to avoid cutting trees by girdling them and how to avoid plowing through the planting of Indian corn in hills rather than furrows. Over the long run, trees were cut down, and as settlement moved westward, the prairies were plowed. Such work was enormously expensive in terms of time and effort. It has been estimated that a farmer in a lifetime of work could hardly hope to clear more than 200 acres of land, but by 1850 perhaps 100 million acres had been made ready for the plow.

There was always a shortage of labor on American farms, and the gulf between the task to be done and the hands available to accomplish it led here, as in the mills and factories of the nation, to thoughts of mechanization. The idea of using machines to replace people was a typical one in the new nation, where high labor costs and high prices for products coincided. In the northern states of the Union, between 1820 and 1870, the rural population increased 2.9 times—but during these same years the urban population increased 14.5 times! The proportion of food sent to the cities rather than consumed in the countryside doubled. New technology in the form of canals, turnpikes, and railroads contributed greatly to these new marketing patterns, and the shift from subsistence to commercial farming in turn stimulated the search for new technology.

Agriculture, of course, was then as now not a single undifferentiated activity but a complex and integrated set of tools and techniques, crops, and social arrangements. Wheat, for example, had to be planted on land previously prepared, then cut, thrashed, and finally transported and sold. With the availability of horse-drawn plows, of a new and widespread transportation network, and of a growing commercial market, it was the harvesting (cutting and thrashing) that presented a technological bottleneck to the expansion of wheat culture.

This critical step in the productive process was necessarily seasonal and extremely labor-intensive. First the crop was cut, then prepared for drying, usually by being bundled into sheaves to be stood on end in the open field. When dry, these were gathered and the grain was beaten off the stalk and put into sacks. Bringing in the grain remained what it had been for centuries—a hand process involving cutting with sickles or scythes and separating the grain from the stalks with hand flails. If this process could be mechanized—that is, if a device could be designed which would substitute horse-power for human toil, the acreage per person could be significantly expanded. The man who built the machine and solved the problem was Cyrus Hall McCormick.

Cyrus Hall McCormick

Source: National Archives

McCormick was born on a farm in Virginia on February 15, 1809. His father, Robert McCormick, had for 20 years tinkered with a device to reap wheat but had never been able to hit upon the right design and combination of parts to do the job. In 1831, at the age of twenty-two, Cyrus picked up the task. Rather than modifying his father's machine, he completely redesigned it and combined the seven features of the reaper that today remain standard on all such machines: the divider, the reel, the straight reciprocating knife to cut the stalks, the finger or guards, the platform to catch the stalks, the main wheel and gearing, and the front-side draft traction. His new machine worked reasonably well on local fields during the 1831 harvest, and the following year he ventured to make a public demonstration near Lexington, Kentucky. Still he persisted in making minor improvements until in 1834 he read of similar work being undertaken by Obed Hussey, who was to be a competitor for many years. McCormick warned Hussey of his priority and on June 21, 1834, took out a patent for his reaper. In 1843 he began to license the manufacture of his machine to various firms, but the resulting devices often proved to be poorly made and their failure in the field was frequently attributed to the design itself rather than the workmanship of the particular machine.

In a key decision McCormick decided to manufacture his own reapers, thus insuring a quality control that would maintain the reputation of his machine. In 1847 he erected a factory in Chicago, then only beginning the boom that was to make it the hub of the Middle West. The new location put McCormick in a position to dominate the market of the fast-developing grain belt of the area, and by 1850 he had a flourishing business. Hussey was his first competitor, but after McCormick's basic patent expired in 1848, a horde of new rivals appeared—over a hundred by 1860.

The superiority of McCormick's reaper was revealed dramatically at the first world's fair, the Great Exhibition at the Crystal Palace in London in 1851. At first ridiculed for their paucity and lack of artistry, the American exhibits came eventually to astonish visitors with their practicality and simplicity. On a farm outside London, in wet weather and working a crop not yet ripe, McCormick's reaper beat all other competitors in actual field trials. The London *Times,* no friend of American achievement, generously declared that the introduction of the reaper alone was worth the whole cost of the exhibition. McCormick continued over the next 30 years to win many more major awards in international exhibitions at Paris, Hamburg, Vienna, and other cities.

At home McCormick proved himself at least as accomplished a businessman as he was an inventor. Subsequent improvements and attachments for the reaper were quickly patented and helped maintain his market domination. He pioneered in such business techniques as credit selling and mass advertising, and in his factory he took advantage of the developing techniques of mass production. By these and other methods he was able to dominate, if not actually monopolize, the reaper market for more than two generations.

McCormick's success was crowned with great wealth and international recognition. He was made an officer of the French Legion of Honor and a member of the Academy of Sciences. Much of his fortune was invested in western gold and silver mines and railroads, and he was aggressive in seeking to promote international

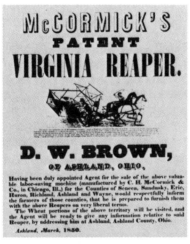

A handbill distributed about 1850 advertises McCormick's reaper and illustrates his advertising acumen.

trade for the Mississippi Valley. Some of his fortune went into the endowment of religious institutions, both newspapers and seminaries. In politics he was a Democrat, but his influence suffered because of the decline in the fortunes of that party due to the Civil War.

Despite its enormous success the McCormick reaper solved only part of the problem of harvesting. The arduous task of thrashing was left untouched by the original machine, and it now became the focus of inventive activity. McCormick himself attacked the problem and beginning in the 1860's produced a number of devices to attach to his reaper which would bind sheaves automatically. Many of the earliest thrashing machines were stationary, steam or horse-powered, and were positioned in the fields where the cut wheat or other grain was brought to them and the grain sacked and carted away.

When the thrasher was combined on a single moving platform with the reaper, the resulting device was known as a combine. During the last decades of the 19th century the vast wheatlands of the Middle Border states (between Texas on the south and North Dakota along the Canadian border) and the central valleys of California saw the development and use of huge combines to cope with the enormous farms that were established in these areas. Machines pulled by 30 or 40 horses or mules cut a swath 18 feet

Teams of mules and horses pull large combines in late 19th-century wheat fields.

Source: U.S. Department of Agriculture

wide, thrashing, cleaning, and sacking the grain as it moved along. By the end of the century even larger combines were being pulled by steam traction engines, the first of which had appeared in 1882, just two years before McCormick's death. In 1898 one writer described seeing one "monster of California farming" used for "cutting, thrashing, cleaning, and putting in sacks at the rate of three sacks per minute of barley, each sack weighing one hundred and fifteen pounds, requiring two expert sack sewers to take the grain away from the spout, sew the sacks, and dump them on the ground. Seven men constitute the crew, including engineer and fireman."

The great increase of land put to the plow—the acreage of wheat alone doubled between 1866 and 1878—put a large premium on scale. More power was wanted to work larger machines. As for the past 200 years, efficiency and a careful husbanding of resources was thought less important than speed and size. By 1890 some 2 million horsepower were being developed by steam on American farms, and by 1910 that figure had reached its peak of 3.6 million. Weighing sometimes 20 tons (but rated at only 50–100 horsepower) and consuming 3 tons of coal and 3,000 gallons of water per day, these monster machines soon gave way to the cheaper and lighter gasoline-powered tractors. By 1923 there was an average of 4.74 horsepower available for work on each American farm. Of this amount 17 percent was provided by gasoline tractors, trucks produced 15 percent, stationary gasoline engines 14 percent, steam engines only 6 percent, electric motors 5 percent, windmills 1 percent, and animals still accounted for the largest share, 42 percent. The steady adoption of gasoline tractors and, after 1935 when the U.S. Rural Electrification Administration was established, electricity for the farm effectively provided American farmers for the first time with power cheap enough and flexible enough to allow a true industrialization of agricultural production.

The appearances of McCormick's reaper, then the combine, and large steam engines to pull them, were dramatic and effective steps toward farm mechanization, but for many years they were far from typical. Such machines were so expensive and could work so many acres that only well-to-do farmers with large acreages could afford them or use them efficiently. The adoption of machines and prime movers varied greatly from place to place and from crop to crop. Whole types of farming—such as raising vegetables and fruits, running dairies, and raising chickens—had little use for large steam traction engines and none at all for reapers. Despite untold numbers of ingenious devices from the imaginations of hundreds of inventors, the basic productive unit remained a human being using traditional hand tools. Even in those areas of agriculture where power and machinery were most successful—gathering grain and grass and tilling the soil—the vast majority of farmers, especially in the northeastern and southern states, were either too poor to afford the new machines or had farms too small, too diversified, too rocky, or too steeply laid out on hillsides to use them.

For most American farmers, therefore, the invention of the reaper was not as important as the far less dramatic improvements in the production of hand tools. By 1850 the American System of manufactures, begun by men like Eli Whitney and John H. Hall in federal and private small-arms manufactories, had spread to the fabrication of agricultural implements. In 1849 a Scottish chemist touring the country attended an agricultural fair at Syracuse, New York and commented that "the general character of the implements, was economy in construction and in price, and the exhibition was large and interesting.... Ploughs, hay-rakes, forks, scythes, and cooking-stoves were very abundant, and many of them well and beautifully made. American ploughs are now exported in considerable numbers.... The potato grips and forks, of various kinds, cut out of sheet-steel, were very elastic, light, strong, and cheap." Significantly he also remarked that "the cradle-scythes were also excellent: an active man was said to be able to cut four to six acres of wheat a-day with them...." He heard news of McCormick's reaper and remarked that the "machine, I suppose, must be a good one, from the information here given that as many as fifteen hundred of them have been made at Chicago ... this year, and sold for cutting wheat on the prairies of the North-Western States," but the only machine at the fair was one of Hussey's, which people claimed could cut 25-acres of wheat a day.

The characterization of agricultural implements as "very elastic, light, strong, and cheap" indicates that they were typical of American technology during these years of the mid-19th century. Two visiting British machine-makers noted that "templates and labour-saving tools are used in the manufacture of these implements, which are sold in very large numbers." Plows thus mass-produced by labor-saving machines sold for as little as two or three dollars, making them available to farmers in large numbers.

In a period when most Americans lived in rural areas (it was not until 1920 that a majority of Americans lived in cities), these mass-produced agricultural implements represented a democratic technology at its very best. It has been estimated that in 1862 it cost only $968 to equip a medium-size general farm, and thanks in part to the spreading techniques of mass production, that cost had fallen to an estimated $785 by 1907. Along with free government land, a cheap, effective, and readily available technology continued to give support to the dream that the nation could remain what it had always been, a country of small, independent, self-reliant agriculturalists.

But that dream, whatever its visionary power, was already doomed. The mechanization so mightily symbolized by the reaper of Cyrus Hall McCormick—with its premium on size and power and consequent demand for large new sources of energy—was already well under way and increasingly embraced by those who were well enough placed to take advantage of it and turn it to their own profit. On the one hand, this mechanization allowed fewer workers to feed more people and thus underwrote the movement toward urbanization and industrialization. But on farms themselves the breaking of the harvest bottleneck created new bottlenecks at other points in the productive chain and thus put a premium on still further mechanization. Farmers who before had been able to plow more than they could harvest could now harvest more than they could transport to market. Large-scale farms, worked by large machines and dependent upon huge grain elevators and railroad networks as well as commodity markets, increased the risks and rewards of agriculture. Nonrenewable energy sources of coal and gasoline replaced those of wind and animal muscle. Gradually agriculture began to demonstrate those same characteristics of mechanized, integrated, energy-intensive production that were also coming to mark most manufacturing industries. It was a result that McCormick himself hardly foresaw and for which he was not alone responsible. But the reaper, more than any other machine of its century, showed what power and mechanization could do for agriculture.

8

JAMES BUCHANAN EADS:
THE ENGINEER AS ENTREPRENEUR

John A. Kouwenhoven

Engineers and architects are grouped together as one of the 16 classes eligible for election to the Hall of Fame for Great Americans—a national institution established in 1900 and administered by the Senate of New York University. Only one person out of the 89 men and women elected to the Hall during its first 60 years of existence was chosen in this category: James Buchanan Eads, engineer, who was elected in 1920 —33 years after his death.

Three years before he died in 1887, Eads had been the first American to receive from Britain's Royal Society of Arts the Albert Medal—awarded in previous years to such distinguished Europeans as Faraday, Bessemer, Lord Kelvin, and Pasteur—and again it was for "the services he had rendered to the art of engineering" that he was honored.

Yet in 1871, when Eads was at the height of his engineering career, he was classed as a "Capitalist," along with Daniel Drew, Cornelius Vanderbilt, and Cyrus W. Field, in a popular book of biographical sketches entitled *Great Fortunes and How They were Made*, by James D. McCabe, Jr., a highly competent journalist and historical writer. And two years after Eads died, Captain E. W. Gould, the garrulous historian of river navigation in America, whose *Fifty Years on the Mississippi* appeared in 1889 and who had known Eads well for many years, claimed that whatever credits were due Eads "as an engineer, or for his mechanical and inventive genius, all sink into insignificance when compared to his ability as a financier."

Captain Gould's judgment was colored, no doubt, by the fact that he had become a large stockholder in one of Eads's ventures and was left with considerable losses after Eads left. Eads had formed a corporation in 1855 to take over the salvage business which had made him wealthy and from which he finally retired in 1857 for reasons of health. Gould, a director and for a while president of the company, preferred to believe that it was Eads's "financial genius" which prompted him to dispose of the company to his friends "just at the moment when the tide of its success was about to turn... when all that was known of (the) value (of its stock) was from representations of interested parties," rather than that it was Gould's own bad judgment or bad luck that caused his losses.

Yet Captain Gould was nearer the truth about Eads than the authors of two book-length biographies when Gould insisted that Eads's two great engineering achievements—the steel-arch bridge across the Mississippi at St. Louis which bears his name and the jetties that secured a deep channel for shipping at the mouth of the river below New Orleans—were monuments to his financial ability as well as to his public spirit and his mechanical and engineering genius.

We can make sense of Eads's career as an engineer only if we recognize—as these biographers did not—that he was also a banker, a railroad promoter, and one of the creative entrepreneurs who shaped our industrial civilization by finding ways to organize and develop the potentialities of modern technology.

* * *

In Lawrenceburg, Indiana, where he was born in 1820, and later in Cincinnati, Ohio and Louisville, Kentucky, where his father took the family in a series of never-successful attempts to better his fortunes, young Eads displayed the fondness for mechanical things that many boys in western river towns developed in the early steamboat era. By the time he was 11, he had made a small engine that ran by steam, and after the family moved to St. Louis, Missouri in 1833, he had a chance to supplement his native mechanical ingenuity with spare-time reading. He had the use of the private library of a merchant for whom he ran errands and did chores in order to supplement his family's income from the boarding house his mother ran. Except for a few years of schooling back in Cincinnati and Louisville, Eads was entirely self-educated.

By the time he was 19, his father and mother had moved once more, this time to a new settlement up river in Iowa, and Eads got his first job on the river, as second clerk (or mud-clerk) on Captain Gould's *Knickerbocker,* running between Cincinnati and Dubuque, Iowa. His mechanical skill had developed to the point where he had designed and built what Captain Gould described as "a miniature steamboat... ready to raise steam in a tin boiler, ingeniously and systematically arranged." But his job as mud-clerk had nothing to do with mechanics; the clerks on river boats were business managers who dealt with passengers and shippers and hired and paid the crew.

Eads quickly demonstrated that he was as talented in dealing with people as in mechanical contrivances. When boats were overcrowded, as they often were, or delayed by low water or accident, it was the clerks who were responsible for placating disgruntled passengers and crews. Captain Gould remembered young Eads as one who, "with the suavity that characterized and popularized him in later years, did much to relieve the captain and quiet the irritation, especially in the ladies' cabin."

The *Knickerbocker,* like hundreds of other boats on the Mississippi, was sunk by a snag in 1839. Captain Gould and his clerks managed to get all the passengers and crew safely ashore, but the valuable cargo, including a large quantity of lead from the mines of Galena, Illinois was lost. River pirates soon made off with the light deck freight, but the tons of lead remained in the sunken hull since there were at that time no wrecking boats equipped to recover it.

The lesson was not lost upon Eads. While serving as a clerk on other boats during the next two years, he developed designs for a diving-bell capable of recovering the machinery and cargo of wrecked steamboats. Early in 1842 he formed a partnership with the proprietors of the first boat-building yard in St. Louis (established in November 1841) to construct his "Submarine," as he called it. He was ready to go into the salvage (or wrecking) business.

Eads was the active, practical man in the firm, in charge of the boat and the salvaging operations (hence the title Captain, which thereafter clung to him) and spent much of his time for the next two years working in his diving-bells on wrecks up and down the river. In 1845, however, he married and wanted to be at home in St. Louis, so he left the salvage firm and turned his attention to building and managing the first glass works west of the Ohio River. His unpublished letters, written during the next few years, make clear that he rapidly mastered the complex technology of flint-glass manufacture, learning how to build the melting pots of German clay, mix the sand, pearl-ash, lead, and other ingredients, and draw off the molten glass into the pressing moulds. But the expenses were great, and by the end of 1847 the glassworks was shut down. Burdened with debts, Eads went back into the salvaging business with his former partner.

The design of Eads's first "Submarine"—not submersible itself—had been basically an adaptation of the twin-hull snag-boats built in the early 1830's by Henry Miller Shreve. But Eads equipped his boat with derricks to lower and raise diving-bells of his own design and with the machinery necessary to propel the boat, operate the derricks, and supply compressed air to the diving-bells. It had worked well, but when Eads returned to the business, he at once began designing and building a series of improved "Submarines." The fourth, built in 1851 at a boatyard in Paducah, Kentucky, revolutionized the wrecking business, for it was equipped with powerful steam-driven centrifugal pumps designed by Eads under an agreement with the patentee (J. Stuart Gwynne of New York) specifically to clear wrecked steamboats of the heavy load of sand and silt deposited in and upon them by the muddy river. Fitted out with these pumps and with more powerful derricks, *Submarine No. 4* was able to raise wrecked steamboats entirely and quickly restore them to service instead of merely salvaging their cargoes.

In 1855 Eads and his partner William Nelson bought five Shreve snag-boats offered for sale by the U.S. government and converted them for use as diving-bell boats—one of which, *Submarine No. 7,* the largest and most powerful boat of the kind ever constructed, figured prominently in the next phase of Eads's career during the Civil War.

The government's sale of its snag-boats resulted from the decline in federal funds for internal improvements during the period of bitter sectional controversy prior to the war. The work of removing snags and other obstructions from western rivers had almost ceased by 1853, and by 1856 the river commerce of the Mississippi Valley was seriously threatened by the hazards of navigation and the inflated costs of insurance on boats and cargoes. In that year Eads got his friends Congressman Luther M. Kennett and Senator Henry Gever to introduce bills in the U.S. Congress for the improvement of all the major western rivers on a contract basis. At the same time Eads organized a corporation, the Western River Improvement and Wrecking Company, to take over his salvage business. (This was the corporation in which Captain Gould later regretted investing.) It was to bid, as no other organization was equipped to do, for the government contract when Congress passed the bills his friends had submitted. Early in 1857 the House passed its bill, but the Senate version died. Similar bills introduced in 1858 were defeated, largely through the influence of Henry Clay and Secretary of War Jefferson Davis.

* * *

Meanwhile, Eads had suffered a severe recurrence of the tuberculosis which periodically incapacitated him throughout his adult life, and—as indicated earlier—had retired from active work. For the next few years he and his second wife lived hospitably in a large Italianate villa in suburban St. Louis, and he invested his considerable wealth in real estate and street railways and became a prominent director of one of the St. Louis banks. But even before the outbreak of the Civil War, Eads had worked out a plan for blockading the Mississippi south of Cairo and defending the rivers with ironclad gunboats. Early in 1861 he was called to Washington for consultations with the Secretaries of War and the Navy.

Eads's initial contract with government was to build seven ironclad gunboats in 65 days. The boats were not, in fact, completed on schedule, nor did Eads have much to do with their design beyond supplying rough sketches to Samuel Pook, the naval constructor at the Washington Navy Yard who drew up the plans. The delays in their construction were the result of the beleaguered government's failure to make payments as they became due and of repeated changes in design ordered by the military authorities. That all seven were completed in 100 days (during which Eads also converted his *Submarine No. 7* into the ironclad *Benton* to become the flagship of the gunboat fleet), was a triumphant demonstration of Eads's organizational skill. Within two weeks of the signing of his contract he had 4,000 men at work in iron mills, machine shops, and mines from Pennsylvania and Ohio to Wisconsin and Michigan, as well as in the foundry and shipyard he had built in Carondelet, just south of St. Louis, where the keels of four of the ironclads had been laid. The vessels were still unpaid for when they helped capture two Confederate strongholds on the Tennessee and Cumberland Rivers—Forts Henry and Donelson—which resulted in the first major Union victories of the war.

While building the ironclads, Eads was thinking about more effective warships for use on the western rivers. Less than a month after the classic 1862 battle between Ericsson's ironclad *Monitor* and the Confederate *Merrimack*, he was called to Washington by Navy Secretary Welles and asked to design all-iron gunboats of very light draft. In the office of the Chief Naval Constructor he borrowed a drawing board and sketched a design for a single-turreted monitor, turtle-backed up from an extremely low freeboard. Eads proposed a turret of his own design, revolving on ball bearings instead of on a central spindle as in Ericsson's patent, and with gunports that closed automatically when the guns recoiled. At first the Navy Department resisted Eads's design for the turrets, and the first three iron river-monitors he built at his greatly expanded shipyard and iron-works in Carondelet were equipped with Ericsson turrets. But he was subsequently given a contract to build four double-turreted river monitors, two of which were equipped with one Ericsson turret and one of his own design (which became the standard in the future), with its guns manipulated entirely by steam. It was these monitors that served Admiral Farragut so well at the battle of Mobile Bay.

After the war Eads continued to design improved gun carriages, turrets, and war vessels. In 1868, after spending a year abroad as a special representative of the Secretary of the Navy, he submitted a report to the department advocating a comprehensive "System of Naval Defenses." His gun carriage, "run out and otherwise regulated by steam," was exhibited by the Naval Ordnance Bureau in the Government Building at the Centennial Exhibition in 1876, side by side with John Ericsson's, which "worked by hand power, taking the united effort of four men to direct its movements." But Eads's contributions to the development of naval armaments were always derogated by the irascible Ericsson (who declined election as a fellow of the American Association for the Advancement of Science because Eads was similarly honored the same year), and they have been largely overlooked by historians. However, his familiarity with the technology of metal working and the properties of iron and steel, as demonstrated by this naval work, helps to explain his next great achievement.

* * *

Eads had no formal engineering training, nor did he have any experience that seemed, on the surface, to qualify him as a civil engineer capable of designing and supervising the construction of even an ordinary bridge. Yet between 1867 and 1874 he spanned the Mississippi at St. Louis with a structure of unprecedented boldness. Eads Bridge, as it is named—the only major bridge named for its chief engineer—has justly been called "in many ways the most remarkable engineering job in bridge history." Its masonry piers were founded on bedrock at greater depths than men had ever worked before (or ever worked again for many years), beneath the waters of a river whose violence in season of high water and when ice-jams formed in winter posed problems no earlier bridge-builder had encountered. Its arches—two of 502 feet (153 meters) and one in the center of 520 feet (160 meters)—were more than 60 meters longer than any ever built before. It was the first large structure ever erected whose principal structural members (the tubular ribs of its arches) were of steel—a "new" structural material banned for use in bridges by the British Board of Trade until 1877. (It employed not just the Bessemer steel which first became commercially available in the United States in 1865, but a metallurgically advanced chrome-alloy steel in whose development Eads had played a part.) The novel method employed in the bridge's erection—cantilevering segments of the arches out from adjacent piers until the steel tubes met at the center, thereby avoiding the use of the supporting "falsework" that traditionally obstructed navigation while arches were being built—established a precedent that has been followed in the construction of large metal arches ever since. It also led to the design of the first American railroad bridge of the true cantilever type: the Kentucky River Bridge (1876) by C. Shaler Smith, a young engineer who had been in St. Louis during the erection of Eads Bridge to design the railroad trestle for its eastern approaches.

Eads Bridge, finished in 1874, pioneered new materials and methods and still carries trains and traffic over the Mississippi River at St. Louis.
Source: St. Louis Chamber of Commerce.

Eads was ably assisted by some professionally trained engineers on his staff—including Henry Flad, who planned the method of erection, and Charles Pfeifer, who did the mathematical computation involved in designing the arches, both of whom had been educated and trained in Germany. However, Eads was personally responsible for the overall conception of the double-decked, triple-arched bridge, for the detailed design of the pneumatic caissons used in founding its piers on bedrock, and for the details of its ribbed-arch superstructure. One of the German-trained engineers who worked in Eads's drafting room and was proud of the contributions of his fellow Germans on the engineering staff nevertheless insisted, years later, that Eads deserved "exclusive credit" for designing and proportioning the bridge. "All the minutest details of the tubes with their couplings and pin connections, the skewbacks and anchor bolts, all of it to the last ⅛ of an inch was the work of Eads," he told an audience of engineers.

Eads's improvements on the pioneering European developments

in pneumatic caissons included several patented inventions, among them the sand-pump, which greatly advanced the technology of underwater excavation. He had familiarized himself with the most advanced European practice before designing his caissons, but unlike other American engineers who contributed to the development of the new technology, Eads had first-hand knowledge of the technical and human problems of working in compressed air, acquired in his diving-bells (which are in essence small caissons) during years of salvage work on the bed of the Mississippi. And his courageous determination to use steel as the basic material in his ribbed arches was grounded in first-hand metallurgical experience gained while working with the naval ordnance people during and after the war.

For several months, just after actual construction of the bridge had begun, Eads was so sick that he could not work, and W. Milnor Roberts, an engineer of international reputation and vast experience, temporarily took charge of the project. Said Roberts in a personal letter to Eads after he resumed his duties: "The Bridge, in its inception, in its plan, and in its noble battle against very fierce and extreme opposition, is eminently yours; and on every account you should control every part of its plans and arrangements." But, as Roberts also said, Eads as chief engineer was in position different from chief engineers in general "not merely from being *projector* as well as the designer of the work" but also because he was "one of the largest owners, and one who has induced the subscriptions, and who is justly looked up to by all the stockholders, and held by them especially responsible."

The story of the financing of the project, and of the "fierce and extreme opposition" Eads encountered, as told in Calvin M. Woodward's *History of the St. Louis Bridge* (1881), fully justifies the description of Eads as a creative person who combined the talents of a great engineer with those of a great entrepreneur. Yet not even in Woodward can one find an answer to the question: Why did a man who had never built a bridge before and who had spent most of his life on the river (the role of which as a highway of commerce was threatened by railroads thrusting across its narrower reaches upstream) suddenly marshall all his varied talents to build a railroad and vehicular bridge at St. Louis, where the combined waters of the Upper Mississippi and the two thousand-mile-long Missouri River presented such formidable hazards?

The oversimplified answer is that Eads had long been involved in several enterprises of great magnitude which his biographers have completely overlooked. When he began planning the bridge, he was the controlling stockholder in the North Missouri Railroad, with one branch building westward to Kansas City and the other northward to the grain-producing regions of Iowa. As Eads and his associates conceived it, the northern branch would secure to St. Louis the traffic in grain that had formerly come down the rivers to St. Louis but was then being reached by railroads thrust west from Chicago and transported on the Great Lakes waterway to the East. The western branch to Kansas City in 1867 seemed to have as good a chance as any of Chicago's lines of becoming a principal link in the first transcontinental railroad system—*if* a bridge at St. Louis were to join it to the track of the Pennsylvania Railroads' western extensions ending at East St. Louis. (Tom Scott, the empire-building vice-president of the Pennsylvania Railroad, was on the original board of directors of the company Eads formed to build the bridge, and five of the other eight directors were associates of Eads on the board of the North Missouri Railroad.) In order to finance the completion of the railroad and the preliminary work on the bridge, Eads headed a syndicate to buy the State Bank of Missouri—the largest bank in the West—which he converted into a national bank with a million-dollar loan from the nation's largest bank in New York—a loan he secured on his personal note without putting up any collateral.

The financial panic of 1873 and the subsequent depression, together with costly delays in completing the bridge caused by political opposition, by mechanical and metallurgical problems inherent in developing new technologies, and by Eads's refusal to reduce his standards of quality, made it impossible for Eads to carry his grand railroad scheme to completion. Several years passed after completion of the bridge in 1874 before the tolls received equalled the interest on the bonded indebtedness of the bridge company. The bridge was sold when the bondholders fore-closed their mortgages upon it. Meanwhile the North Missouri Railroad had also been sold under foreclosure, and in 1877 Eads's bank closed its doors in the largest national bank failure up to that time. Two years later in 1879, both the bridge and the railroad came under control of financier Jay Gould as keys to his southwestern railroad empire.

By that time, however, Eads had successfully completed the largest project in hydraulic engineering ever undertaken by a private citizen under contract with the United States government: the opening of a deep and permanent channel through the bar at the mouth of the Mississippi by constructing jetties extending the banks of the South Pass two and a quarter miles (3.6 kilometers) out over the bar to the deep water of the Gulf of Mexico. More than a year before his bridge was completed Eads had decided that such jetties were the only practicable way to secure an open mouth for the river. The U.S. Army Engineers and the Chamber of Commerce of New Orleans thought not. Led by Brigadier General A.A. Humphreys, Chief of the Army Engineers, who coauthored the standard work on *The Physics and Hydraulics of the Mississippi*, they "proved that such jetties could not be built at the mouth of the Mississippi and would not provide a permanent channel for seagoing vessels even if they could be." Instead, they proposed that the government should build a canal eastward from near New Orleans to Breton Bay on the Gulf.

Relying on his own first-hand knowledge of the river and on careful study of successful applications of the jetty system at the mouth of the Danube and other European rivers, Eads confidently argued for his jetties before Congressional committees. In March 1875, he finally succeeded in getting a bill through Congress authorizing him to proceed according to his plans. The terms of his contract were extraordinary. Only if the depth of water over the bar had been increased in a wide channel from the normal 8 feet to 20 feet (from 2.44 to 6.09 meters) within 30 months was he to receive any payment whatever. If he did secure such a channel in that time, he would receive $500,000. If he secured two additional feet within 12 months thereafter, he would receive another $500,000, and so on. When a channel 250 feet wide and 30 feet deep had been attained, he would receive a total of $4,250,000, another million coming due over a period of 20 years if the 30-foot channel was maintained.

The $5,250,000 total was considerably less than a Congressionally appointed board of civil and military engineers estimated it would cost to build the jetties and less than half of what a canal would have cost. But Eads had already designed a novel type of "mattress' construction much simpler and more economical than any previously employed and was therefore confident of making a large profit if he could convince potential investors to help finance the years of work required to earn his first payments.

That he succeeded in raising the money despite a savage campaign of adverse publicity by proponents of the Army Engineers' canal project is further evidence of the "financial genius" Captain Gould attributed to Eads. That he also devised a novel system of jetty building and all the machines necessary to carry his plans to an early and successful conclusion is further evidence of his versatile genius as a practical engineer.

Even before the jetties were completed, Eads had begun planning and organizing what turned out to be his last bold project—a ship-railway across the Mexican isthmus of Tehuantepec. This was 1,200 miles closer to the Mississippi's South Pass jetties than Panama, where de Lesseps was laying his disastrous plans to build a tide-level canal. Against the usual opposition of "scientific" and "engineering" authorities, Eads proposed to haul the largest ocean-going vessels overland between the Atlantic and Pacific in huge cradles, drawn on a 12-rail track by two powerful locomotives, each running on six rails. Once again, none of the government guarantees he asked Congress to provide would become operative until he had built enough of the ship-railway to prove that the technological problems had been solved.

A bill to provide a charter and financial aid to his company passed the House in 1880 but died in a Senate committee and for the next few years similar bills were similarly sidetracked. Finally, ill and desperately weary but absolutely convinced that his scheme was a national necessity, Eads told Congress he would finance the entire $75,000,000 project himself, asking nothing of the government but a sanctioning charter for his company.

Just as victory for the charter in both houses of Congress seemed imminent, Eads's health broke, and his doctors ordered him to leave Washington and rest completely. While his wife nursed him in Nassau, his bill passed the Senate, but on March 8, 1887, before it came to a vote in the House of Representatives, he died. His last words, it is said, were: "I cannot die, I have not finished my work."

9

JAMES B. FRANCIS AND THE RISE OF SCIENTIFIC TECHNOLOGY

Edwin T. Layton, Jr.

It has become a truism that we live in an age of science. But the manner in which science influences us has changed. Science used to affect the way men thought; now it affects the way that they live. The work of Copernicus, Galileo, and Newton changed our view of man's place in the universe. But modern science is so abstruse that ordinary people have difficulty understanding it. Thus, while few people understand relativity or quantum theory, everyone is affected by the atomic bomb. The bomb is an example of the interaction of science and technology. Interactions of this sort began in the 19th century, in association with the Industrial Revolution. By the early 20th century technology had become scientific. It was institutionalized in great industrial research laboratories. It produced a flood of inventions that increased the rate of social change and altered the way people live and work. The results have not all been good. On the one hand, scientific technology promises a utopia of material abundance, but on the other hand, it threatens nuclear destruction or environmental catastrophe. People differ on how best to direct technical change for human benefit. But there is little doubt that scientific technology has become one of the most important forces shaping the modern world.

The rapid rise of the United States from a weak agrarian republic in 1800 to a mighty industrial power in 1900 owed much to scientific technology. Faced with the enormous task of settling a virgin continent, Americans turned eagerly to science. The national government fostered the development of scientific technology in several ways. President Thomas Jefferson founded the United States Military Academy, West Point. Its curriculum was strong in science and engineering, and its graduates were encouraged to become civilian engineers. America's immigration policies also contributed to the development of scientific technology. Among the millions who migrated from Europe to find a better life in the New World, there were a number of highly trained engineers. Engineers, both native-born and immigrant, were in the vanguard of those who attempted to base technology upon science. They produced many innovations that contributed to America's rapid growth. Among the most important of these men was a young British immigrant, James Bicheno Francis. The hydraulic turbine which he invented became one of the most important sources of power for American industry. But the tradition of scientific research in industry, which he began, has had an even greater influence.

Francis was born at Southleigh, England. He was trained as an engineer by his father, John Francis, who built an early railroad in his native land. After a brief education at Wantage Academy, young Francis went to work, first for his father, and then for a succession of canal companies, in the course of which he became thoroughly grounded in the engineering practice of his day. In 1833, at the age of only 18, he migrated to America. Competent engineers were scarce in the New World, and Francis became the assistant to one of America's leading engineers, Major George W. Whistler, a graduate of West Point. In the following year Whistler was called to the industrial city of Lowell, Massachusetts, one of America's first cotton textile manufacturing centers. Francis accompanied Whistler to Lowell and continued to serve as his assistant until 1837 when Whistler left. The Proprietors of Locks and Canals of Lowell, the corporation that developed the city of Lowell and its water power, then made Francis their chief engineer. Thus, after only four years in America and at the age of only 22, Francis assumed one of the most important and responsible engineering positions that America had to offer; for Francis the

American dream of opportunity was fully realized.

Francis quickly vindicated the trust placed in him. His chief duties were to maintain and rebuild a system of dams and canals which provided the water power that drove the factories at Lowell. He also had to measure the flow of water to the various cotton mills and to advise the factory owners in all engineering matters. In his characteristic manner, Francis patiently gathered data first and then sought rational, scientific solutions to the problems. He found that a half-century earlier there had been a flood before the industrial complex at Lowell had been built. If the water were to rise again to the same height, it would flow over the main canal lock and do a great deal of damage to the new industrial city. Francis therefore increased the height of the side walls of the canal and built a massive wooden gate 25 feet high which could be lowered across the mouth of the canal. Within two years another flood occurred; without Francis's foresight a large part of Lowell and its factories would have been destroyed, with the probable loss of many lives. As it was, no damage was done, thanks to Francis's scientific approach to his job and his good judgment.

Francis depended upon the existing scientific literature in engineering, which at this time was mainly European. In some cases—perhaps the majority—this was all that was needed. Following British practice, the factories at Lowell were built on a framework of iron beams and pillars. Francis relied upon British data on the strength of iron in evaluating the soundness of plans for new structures at Lowell. But he soon found that in this and other cases the existing scientific knowledge was insufficient, and he had to supplement it by conducting his own experiments. Thus in attempting to use science as the basis of engineering practice, Francis was driven to become a scientist in his own right. This proved to be a congenial role, for he had both keen insight and a flair for experimental investigation.

Francis's career provides an example of how technology became scientific. Engineers borrowed the methods of science, as Francis did in his experiment to determine the strength of iron beams. In this way engineering knowledge, like that of science, became cumulative. Each man's work could build on that of his predecessors. Another trait of science was abstraction, exemplified by its mathematical theory. When applied to technology, this sometimes led to sudden, dramatic changes. The traditional approach to tech-

94

nology had been to make small changes in existing devices. But once engineers understood the theory underlying a device, it was sometimes possible to make rather radical improvements. A case in point was the invention of the hydraulic turbine.

For centuries water wheels had been the basic source of industrial power, but the conventional water wheel had a practical limit in its efficiency. At best only about two-thirds of the energy of falling water could be transformed into mechanical energy to drive machinery. By using mathematical theory the French engineer Benoit Fourneyron produced a radical improvement in the 1820's, the hydraulic turbine. It increased the efficiency of water motors to nearly 80 percent. Francis then improved significantly upon Fourneyron's design; indeed, the Francis turbine is still the most widely used type. But Francis also made improvements in the science applied to the turbine which were of equal or greater importance.

To understand what Francis accomplished, it is necessary first to understand what Fourneyron did. France had been a pioneer in atttempting to base engineering upon science. French engineers, notably Jean Charles Borda and Lazare Carnot, had shown the major limitations of traditional water wheels, and they had enunciated the conditions for greater efficiency. They found that the energy of the water was lost in two ways: by ''shock'' at entrance and by excess velocity at exit. By ''shock'' they meant the turbulent motion of the water produced when a stream of water struck the paddle of a waterwheel. Thus the conditions for highest efficiency were that the water should enter the wheel without shock and leave it with only the minimum velocity needed to clear the wheel.

Fourneyron's achievement was to show how these conditions could be approximated in practice. His turbine was much like a conventional waterwheel turned on its side. The water entered at the center and flowed outward in a horizontal direction. The wheel was enclosed by metal plates to prevent the water from leaking out, and the blades of the wheel were curved. By curving the blades Fourneyron showed how to achieve the second condition: exit with minimum water velocity. His analysis showed that this condition would be achieved when the water left in a direction precisely opposite to the wheel's motion, that is, in the direction of a tangent drawn to the circumference of the wheel. Fourneyron could approximate this condition by giving the blades of his wheel just the right amount of curvature.

Shockless entry was the hardest condition to meet. This implied that there would be no impact of the water upon the wheel. Fourneyron's theory indicated how this could be achieved. The water should enter in a direction exactly tangent to the moving blade; since the water would be moving parallel to the axis of the blade there would be no impact and consequently no loss through "shock" or turbulence. To meet this condition, it was necessary to control the angle at which the water entered the wheel. Fourneyron approximated this by making his wheel hollow and mounting fixed guide vanes inside the moving wheel. These vanes directed the water to the wheel at the required angle.

Fourneyron's turbine represented a remarkable advance. But it was considerably more expensive than traditional waterwheels, and its use was therefore limited to great industrial centers where there was plenty of capital and a pressing need for higher efficiency. A cheaper turbine would be of enormous benefit, particularly in America where water power was abundant but capital was scarce. The Francis turbine, when fully developed, provided the ideal answer; it was much cheaper than Fourneyron turbines and usually got significantly higher efficiencies. Francis's improvements, like Fourneyron's own work, were based on a scientific analysis of the problem. In effect, he expanded the science of the turbine. But he did more, for the defects he saw in the theory of the turbine were in fact quite general: they applied to virtually all attempts to apply science to technology.

There were two basic defects in existing scientific technology. First, the mathematical theory was too idealized or simplified. It left out such things as friction which were of vital importance. Second, attempt to apply experiments to machines had run into perplexing problems due to "scale effects." The results were often not consistent, varying in unpredictable ways as the size of the machine changed. Francis did not solve either of these problems completely, but he found ways of circumventing both limitations. Thus Francis showed not only how to improve the turbine but how to create a more scientific technology.

In this work Francis was not alone. He collaborated with another engineer, Uriah A. Boyden. It was Boyden who first introduced the Fourneyron turbine to Lowell in 1844, and he patented a number of improvements. A few years later he sold his patent rights to the Proprietors of Locks and Canals of Lowell, and he then collaborated with Francis over many years in improving the design of turbines and the methods used in testing them.

The talents of the two men were complementary. Boyden was the abler at mathematical theory, while Francis was the greater experimentalist. Boyden was a shy genius, almost a recluse: he published none of his findings and had no close professional associates apart from Francis. Francis, who was warm and gregarious by nature, published the results of their collaboration and trained a generation of younger engineers in their methods. Boyden had a profound understanding of the defects of the existing hydraulic science, and he did much to set Francis upon the right course. But Francis had the experimental skills to show how these deficiencies might be overcome. The insights of Boyden and Francis led each of them to see how a radical improvement in turbine design could be made. But Francis was apparently first, and he was the one to publish his results. Thus the Francis turbine rightfully bears his name. On these foundations, both scientific and practical, an enduring tradition of industrial research grew up in America which drew inspiration from Francis.

To understand the work of Francis and Boyden it is necessary to understand the theoretical and practical problems which they surmounted. The existing mathematical theory was beautiful and elegant, but it was too idealized. This defect was a matter of necessity, however. The most important factors omitted from the theory were friction and the internal resistance of fluid. Not until the 20th century were satisfactory theories evolved to understand these phenomena.* Boyden saw this problem with great clarity, and he stressed the limitations of idealized theory in his correspondence with Francis. But Francis fully grasped the point and used it. The core of hydraulic science lay in measurements of the flow of water. But none of the formulae available were entirely

*Both friction and water resistance represent a loss of the energy available to do work. Water resistance arises when some of the particles of water move in a direction different than that of the fluid flow. If the motion of the particles is circular we speak of eddys; if the motion is random, in all directions, we speak of turbulence. While we still lack a comprehensive theory of turbulence, many of the phenomena of fluid motion which perplexed 19th-century engineers have been clarified by the development of the science of fluid mechanics in the 20th century.

accurate. As Francis wrote after reviewing existing theories:

> ...the result, however, of these numerous labors is far from
> satisfactory to the practical engineer. On a careful review of
> all that has been done, he finds that the rules given for his use,
> are founded on the single natural law governing the velocity of
> fluids, known as the theorem of Torricelli; omitting in conse-
> quence of the extreme complexity of the subject, all consider-
> ations of many other circumstances, which it is well known,
> materially affect the flow of water though orifices.

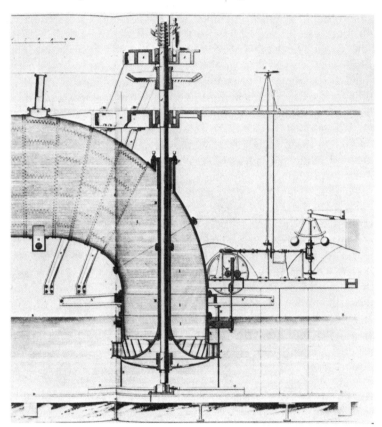

*The Boyden outward-flow turbine is Boyden's modification of the Fourneyron
design. The water enters fixed guide vanes at L and moves outward through curved
buckets.*

Source: Francis, Lowell Hydraulic Experiments, *Plate I.*

The existing mathematical theories of the turbine, such as Fourneyron's, "solved" the problem by a convenient assumption: that friction and fluid resistance could be omitted. This was a necessary assumption at that time, if any mathematical theory was to be constructed. But Fourneyron and his successors converted a mathematical necessity into a principle of design. They assumed that the actual path of the water was of no significance as long as there were no sharp bends to produce turbulence. Boyden realized that this assumption was incorrect; eddies and turbulence were produced by many things, not just by sharp bends. And these were among the most important sources of lost efficiency in turbines. Since the existing mathematical theory could not cope with eddies and turbulence, Boyden's response was to abandon this approach, while continuing to use its key results. He traced the path of a stream of water particles, the "streamline," inch by inch as it moved through the turbine, and by utilizing basic hydraulic and mechanical principles, was able to take additional factors into account and thus reduce losses of efficiency. His turbines got efficiencies of as high as 88 percent. Francis absorbed and expanded upon Boyden's methods, developing graphical methods for designing turbines.

This detailed attention to the actual path of the water flowing through the turbine revealed a fundamental limitation in Fourneyron's design and suggested a possible remedy. In an outward-flow turbine such as that of Fourneyron, the inner circumference of the wheel is of necessity less than the outer circumference. Thus the channels through which the water flows get wider as the water moves through the turbine. This causes the water to spread out to fill the widening channels, producing eddying, a circular motion in the water, leading to a loss of efficiency. That is, an outward-flow turbine, of necessity, loses some of the energy available by converting it into internal motions within the water. There was an obvious remedy: to reverse the flow from outward to inward. This had already been done by an American millwright, Samuel Howd. Howd's wheels had the advantage of being compact and consequently less expensive. And an inward-flow wheel eliminated losses through eddying, since while moving inward the channels got narrower rather than wider. But, as is often the case, when one problem was eliminated another took its place. Narrowing the channel through which a fluid flowed in-

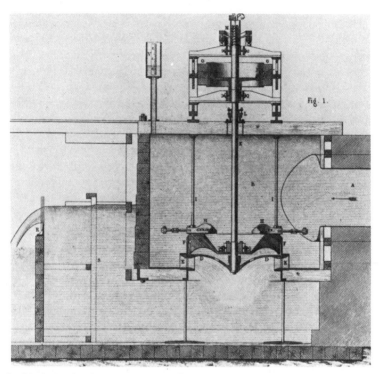

Fig. 1.

In this design for the Francis inward-flow turbine the water enters the fixed guide vanes at E and flows inward and downward through the buckets. A Prony friction brake, part of a Prony dynamometer is shown mounted on the upper end of the shaft at 0.

Source: Francis, Lowell Hydraulic Experiments, *Plate VIII*.

creased its velocity. This was most undesirable, since it meant that the water would have a considerable velocity on its exit from the turbine, representing a very serious loss of efficiency. Thus. Howd's turbine was cheap but quite inefficient.

Francis saw how he could combine the cheapness of Howd's inward-flow turbine with the high efficiency of Fourneyron's outward-flow turbine. The basic problem was to slow the water down. In any inward-flow design the width of the channel must narrow. But Francis saw that the critical fact was not the width, but rather the cross-sectional area of the channel. So that while the width in an inward-flow turbine must narrow, this could be compensated for by increasing the depth, the distance between the top and bottom plates that enclosed the wheel, which were called "crowns." By altering the depth the cross-sectional area—and consequently the velocity—could be controlled. Such a turbine is called a "mixed-flow" turbine, since the water moves not only inward, but downward as well.

Francis mentioned the idea to Boyden who shortly thereafter came up with his own version of the mixed-flow turbine. Francis wrote the following a memorandum on April 15, 1848, probably to preserve his own priority:

> I met Mr. Boyden on the street. I asked him if he did not think that the making of the distance between the crowns of the wheel greater on the inside periphery than on the outside was new . . . . He replied that he had application then pending for patents for this idea . . . . I was much surprised—the more so as I had explained to him some months ago that I intended to make the wheels for the Boott Mill with a greater height between the crowns on the inside periphery than on the outside—and the impression I got at the time was that he thought it would not answer and he gave me not the slightest intimation that he claimed it as his invention.

It is probably impossible to determine whether Boyden or Francis had priority in inventing the mixed-flow turbine. Francis evidently thought he did, and he apparently suspected that Boyden may have taken the idea from him. But it is also possible that Boyden, who was secretive by nature, had been thinking along similar lines and chose not to tell Francis of his work in order to get a prior patent claim. Boyden's own patents on this type turbine represent a rather complex and sophisticated set of ideas, and it is clear that they are not just an imitation. Indeed, they appear to represent the culmination of a great deal of thought over a considerable period of time. It is possible that Boyden's thinking was crystalized by Francis's communication. But there is no question that Francis's work was independent, nor that he was the first to publish his work.

The Francis turbine was not an instant success. The advantages which it seemed to offer in principle were not immediately realized in practice. This is a rather common situation in technology, which necessitates a lengthy and expensive process of development. Before Francis, inventors had to rely principally on intuition and experience in trying to perfect their devices. But in the case of the Francis turbine the development was made more scientific by the systematic use of experimental testing. Thus theories or ideas of how to improve a turbine's performance could be put to rigorous, quantitative test. The basic tool was already at hand, the Prony dynamometer, which had been developed in France and which Fourneyron had used to measure the rotary force, or "torque," developed by his turbines. It consisted of a massive wooden friction brake attached to a balance arm, on the end of which weights could be mounted to measure the torque in a manner quite analogous to the way in which objects are weighed on a scale. But while the Prony brake could measure the torque, or the output, of the turbine, there was no equally satisfactory way of measuring the input, the amount of water used in a given time. Only by comparing

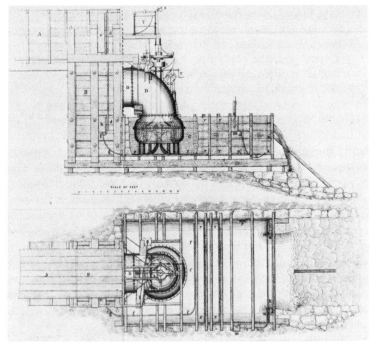

This testing flume with Emerson's dynamometer was constructed by the Swain Company in 1869 and is similar to Emerson's flume in Holyoke, Massachusetts, which became a national laboratory for testing turbines.

Source: Journal of the Franklin Institute, *1870, Plate I.*

output and input precisely could the efficiency of the turbine be accurately determined. The existing formulas for calculating the flow of water were based on small-scale experiments. But here was a case where the results varied with the dimensions of the turbine and of the gate or weir over which the water flowed.

Among Francis's greatest achievements was the development of turbine testing into a matter of exact science. In this, too, he followed Boyden, who perfected several improvements in testing methods and apparatus. To compensate for the scale effects, Boyden and Francis did all their experimentation on full-sized turbines, thus circumventing the unpredictable results of model experiments. But Francis went further and determined a new and more accurate formula for calculating the flow of water over weirs. In this he got some mathematical assistance from Boyden. But the formula was basically empirical, and it was Francis who conducted an exhaustive set of experiments to determine the new formula. Francis then applied it, along with other improvements, to testing turbines with a degree of precision never before attained. Perhaps most important of all, he published his findings in 1855 in his monumental book, *Lowell Hydraulic Experiments*. This book became a guide for American turbine builders in improving and testing the turbine. Francis also trained a number of assistants in his methods, and some of them went on to test turbines. A number of companies conducted their own tests, but in 1870 a former employee at Lowell, James B. Emerson, organized the Holyoke Testing Flume where he conducted public tests of turbines and published the results. It was, in effect, an industrial research laboratory. The result of scientific testing was a spectacular improvement in the performance of Francis turbines and an even more remarkable decline in their cost. By the 1870's variants of the Francis turbine became the most widely used hydraulic prime movers in America, at a time when water power was still more important than steam for industrial purposes. But however important the Francis turbine, Francis's methods and insights were of even greater significance in the long run. Building upon the work of European forerunners and guided by Boyden's insights, Francis showed how a truly scientific technology could be constructed. He developed theoretical and experimental methods that made turbine design more of a science and less of an art. He showed others how to use his methods and insights, and his followers carried on a tradition of industrial research that finds expression in the great industrial research laboratories of the present day. Francis, of course, was not alone. Other engineers in other fields of endeavor were following similar paths. And it was the confluence of these paths that created, for good or ill, modern scientific technology.

10

ALEXANDER GRAHAM BELL AND THE CONQUEST OF SOLITUDE*

Robert V. Bruce

The Industrial Revolution, begun in the 18th century, led in the last half of the 19th century to a communications revolution which shaped the course of industrial and urban growth: modern industrial society could not exist without the instant interchange made possible by developments in communications. For much of the world, the telephone is the symbol of the communications revolution. It not only serves the office and the factory, but it has entered the home to form a major social and personal link between people. Today's telephone systems are global and expanding as fast as installation will permit in most countries. The United States is a nation of telephones with a total of 162 million as of 1977 and with 96 percent of its homes having telephone service.

The search to transmit speech electrically was part of the burst of individual inventive activity which swept 19th-century Europe and America. The invention of the telephone was almost inevitable with the number of minds bent on speeding communications and with the accumulating advances in science, and in the 1870's a

*Adapted by the Forum Editors from Professor Bruce's interview for VOA Forum's radio broadcasts and from his book, *Bell: Alexander Graham Bell and the Conquest of Solitude* (Boston: Little, Brown and Company, 1973).

Scottish-born teacher of deaf-mutes, Alexander Graham Bell, decided to make his fortune by inventing an improved telegraph and came up with the telephone as well.

Appropriately, Bell came from a family preoccupied with speech: his grandfather had taught speech and elocution, and his father, Melville Bell, also made speech his career. At the time of Alexander's birth in 1847, his father was well established in Edinburgh teaching speech, writing and lecturing on elocution, and beginning his research on the workings of the human voice. In his mother, Alexander found another perspective on the human need to communicate. Eliza Bell was almost deaf, and it was Alexander, the second of the Bell sons, who was able to communicate best with her. By speaking in a low voice close to her forehead, he could make his mother hear without her ear-tube. Like most of his family, he had a remarkably expressive, flexible, and resonant voice, but in this peculiarly specialized sort of communication, he seemed better than any of them.

Although he was only a fair student at the Royal High School, the young Bell showed a real aptitude for his father's work. He had a very keen ear, and while not possessed of absolute pitch, he was able to distinguish very fine differences between sounds. Moreover, he was a good musician and even considered becoming a concert pianist.

Edinburgh was in the forefront of mid-19th-century scientific and technological developments. Growing up around the workshops and the mills, Bell gained a knowledge of mechanics, devising a mechanical grain husker, and soon formed an ambition to invent. This, he perceived from the technologically minded men around him, was a way to get ahead. On a challenge from his father, he and his older brother built a speaking machine at home, making gutta-percha replicas of the mouth, throat, and nose, a maneuverable tongue, and bellow lungs, and succeeded in producing human-like cries.

Melville Bell's speech research had as its goal a complete and universally applicable phonetic alphabet: a written system in which symbols represented the positions of the vocal organs and indicated the articulation of all possible sounds. In 1864 Melville Bell completed his "Visible Speech" system, and the three Bell sons gave demonstrations using Visible Speech to pronounce correctly sounds and foreign and dialect words unknown to them.

George Bernard Shaw, a family friend, immortalized Visible Speech in *Pygmalion* with Professor Higgins using it to transcribe Eliza's Cockney dialect. Designed primarily for linguistic scholars and phoneticians, Visible Speech also had practical application in teaching the deaf to articulate, since it showed them graphically how to form sounds they had never heard.

Bell's parents moved to London in 1865, where his younger brother subsequently died of tuberculosis. After teaching in Elgin and Bath, Bell joined the family in 1868 to enter the University of London and help in his father's work. Making use of the potential of Visible Speech, Bell became involved with his first deaf pupils. But before he could complete his studies, his remaining brother died. Already committed to a lecture tour in the United States, Melville and Eliza Bell decided to emigrate to Canada and convinced Alexander to join them. In August 1870, the Bells settled outside of Brantford, Ontario. The following April Alexander found himself, through his father's connections, with a short-term appointment to teach Visible Speech at the Boston School for Deaf Mutes. He found enough pupils to continue the next year, and in 1873 he accepted a professorship of "Vocal Physiology and Elocution" at the recently opened Boston University.

For the rest of his life Bell considered his profession to be a teacher of the deaf and always so identified himself on biographic forms and questionnaires. And in spite of his many later interests and commitments, he always made time to see parents with deaf children, worked energetically for the education of the deaf, contributed to research, and undertook many individual acts of charity to deaf persons. As one to whom Bell gave personal and financial support in her struggle to overcome both deafness and blindness, Helen Keller dedicated *The Story of My Life* to him.

The Boston area in the mid-1800's hummed with scientific and technological activity. The center of industrial New England and the site of Harvard University and the Massachusetts Institute of Technology, the community drew together scientists, engineers, inventors, skilled mechanics and artisans, and businessmen eager to invest their capital. Bell attended many public lectures, read extensively on physics—acoustics and electricity especially—and began a line of work that eventually led him to the concept of transmitting speech electrically.

The scientific principles basic to the telephone had been discovered 45 years or more before Bell invented it. Danish scientist Hans Oersted had demonstrated the principle of electromagnetism in 1820; England's Michael Faraday had published his work on the induction of current in 1831. Oersted discovered that a current passed through a wire coiled around an iron core magnetizes the core. In retrospect, it seems obvious that if the magnet were brought close to a flexible metal diaphragm and a varying current were passed through the coil, the magnet would cause the diaphragm to vibrate to produce sound. To achieve current variations subtle enough to convey speech, Bell was to use a principle of induction: a changing magnetic field will induce or generate a current in a circuit. The mechanical power of the voice could be used, Bell discovered in July 1875, to vibrate a diaphragm fixed near an electromagnet, changing the magnetic field and thus inducing the necessary varying current. He had the original inspiration in the summer of 1874 while visiting his parents at Brantford. Bell was later to joke that the telephone was conceived in Brantford and born in Boston, thereby sidestepping the argument over where the telephone had been invented. His first concept was of a series of tuned reeds rather than diaphragms to transmit and receive the vibrations of the voice, but Bell did not believe that his idea would produce audible sounds.

At that time Bell was at work on two earlier conceptions: a "harmonic telegraph" to send multiple messages simultaneously and a technologically related device he called the "autograph telegraph" to transmit whole documents in facsimile.

Since Samuel F. B. Morse completed his first telegraph line in 1843, telegraphy had burgeoned into a full-fledged technology. Dominated by the Western Union Telegraph Company, this new industry meant instantaneous communication between faraway points that previously had to depend on the physical delivery of messages. The new system, however, still relied on hand delivery between telegraph stations and individuals. Taking their cue from the success of telegraphy, such would-be inventors as young Thomas Edison and Bell looked to improvements in the telegraph to make their fortunes. Bell's telegraphy efforts led him to the telephone and a system which would eventually supplant the telegraph. Ironically, Western Union turned down a chance to buy all of Bell's telephone rights for $100,000 in 1876. Committed to a system incompatible with the telephone, the company shortsightedly missed its opportunity. Another twist of fate later saw the Bell Telephone Company pitted against Western Union in the great patent controversy that developed over the telephone.

Bell drew on his musician's observations for the harmonic telegraph, using the principle of sympathetic vibration—the phenomenon that causes the strings of a piano to resonate selectively, echoing only those notes that are sung into it. He used a set of steel "reeds," each tuned to a different frequency, to open and close electric circuits in vibrating, thus producing intermittent currents of corresponding frequencies. The currents could be sent simultaneously over the same wire to electromagnets bearing upon similarly tuned reeds. Each receiving reed would vibrate selectively in response to the corresponding reed at the transmitting end, thus separating out the frequencies. Bell devised a stylus mechanism to be attached to each receiving reed, which would mark a passing strip of paper whenever the reed sounded.

A model of the first telephone receiver to carry the sounds of speech, set up overnight on June 2–3, 1875, by Bell's assistant, James Watson.

Source: American Telephone and Telegraph Company

Bell's nascent ideas on the telephone were temporarily pushed aside by his work to perfect his harmonic and autograph telegraph devices. He was spurred on by the news that an electrician and inventor, Elisha Gray, was working on a "musical telegraph." But early in June 1875, Bell discovered by accident in Boston that the theory he had conceived in Brantford would work. He found that plucking the reeds of a harmonic telegraph without any transmitter current would induce an undulating current and that the receiver reeds would not only vibrate but sound audibly. Then, using a membrane diaphragm with a metallic center instead of a reed, Bell transmitted the sound of his voice to his assistant Watson, who recognized it but could not make out the words. Bell's next breakthrough came in January 1876. Working on his telephone patent application, Bell added the concept of variable resistance to his specifications, noting that a wire vibrated by the voice while partially immersed in mercury or some other conducting liquid could be placed in a circuit to vary its resistance and produce an undulating current. This meant that transmission no longer had to depend on the limited mechanical power of the voice to induce current, since through variable resistance, an existing current of any strength could be modulated.

But it was not until March 10 that, using a variable resistance model, Bell called out, "Mr. Watson, come here. I want to see you." And Watson in the next room heard.

Meanwhile, Bell had placed a completed patent application in the hands of Gardiner Greene Hubbard, one of his two financial backers and his future father-in-law, who in turn had deposited it informally with the Patent Office in Washington, D.C. For reasons involving negotiations with British investors, Bell had told Hubbard to withhold official filing. Hubbard became impatient, and on the morning of February 14, 1876, without Bell's permission he filed it formally. In a remarkable coincidence, Elisha Gray appeared at the Patent Office several hours later to file a caveat—a statement of a conception not yet reduced to practice—for a telephone device.

It seems probable that Gray had heard of Bell's device to transmit speech electrically, since Gray had been in and out of the Patent Office for several weeks before February 14. At times the most important thing in developing an invention is to know that it is possible. In all likelihood Gray had thus been given the cue to try to do it himself. There was a subtle but fundamental difference between Bell's and Gray's approaches, however, which indicates that neither knew the details of the other's thinking.

There was obviously a conflict, but since Bell had priority in time—although only a few hours—and since he had reached a much more advanced stage of invention, he was granted the patent—No. 174,465—perhaps the single most valuable patent in history.

Western Union subsequently bought out Gray's rights and hired Edison to come up with patentable improvements to the telephone. Edison developed a transmitter superior to Bell's, and with it, a competing telephone company backed by Western Union began to draw customers away from the Bell company. In the ensuing year of litigation it was Bell's patents and Bell's clear and convincing testimony as a witness that saved the day. In settlement Western Union gave up its telephone rights including the Edison transmitter patent for 20 percent of telephone rental income for the 17 years that the patents were valid. The Bell Company emerged with a monopoly which it continued to hold despite some 600 legal challenges over the next two decades.

The immediate struggle Bell faced after receiving the patent, however, was to have people accept his invention. The Centennial Exhibition in Philadelphia presented him with a unique showcase. Bell developed an iron-box receiver in the spring of 1876. It was this device, used with both variable-resistance and electromagnetic transmitters, that he set up in late June in the Exhibition's Main Building to demonstrate to a group of prominent American scientists and—important for publicity—the Brazilian Emperor Dom Pedro. Bell's voice sounding over the device amazed both experts and emperor alike. The success made Bell's telephone the talk of the nation's scientific community, and visiting foreign scientists carried the word abroad. Bell then undertook a series of lecture-demonstrations around the Boston area and in Brantford with improved equipment. Local newspapers soon awoke to the discovery, and Bell became a popular and effective advocate for his brainchild.

At first hearing the telephone was a memorable and even frightening experience for people. Some country audiences almost panicked when they heard a disembodied voice coming from the iron box. But people quickly accepted it. It worked. The telephone quickly became commonplace. The first regular telephone line was established on April 4, 1877, from the shop where telephone equipment was manufactured to the suburban Boston home of the shopowner. In July when the Bell Telephone Company was formed, over 100 telephones were in operation, and by August the number jumped to 600. (At Bell's death in 1922 the company served well over nine million telephones.)

But Bell was not interested in running a business and soon withdrew from the company's board of directors. He gradually ceased to participate in company affairs other than serving as a witness in various patent litigations. By 1881 he had severed his connections with the company, only retaining some 2,000 shares of stock. Bell had previously sold the bulk of the family stock during the boom market for telephone shares. The growth of the business, eventually reorganized into the American Telephone and Telegraph Company, was the result of other men's efforts. Nonetheless, it was Bell who was called upon to inaugurate telephone service between New York and Chicago in 1897—an event captured in one of the best-known of Bell photographs.

For the opening of the Columbian Exhibition on October 18, 1892, Alexander Graham Bell inaugurates the first New York-to-Chicago line.

Source: American Telephone and Telegraph Company

Bell's telephone over the years became an integral part of the American household. Popular legend has it that Bell himself refused to have one in his own home, and, endowed with a rather playful sense of humor, Bell at times encouraged the story. As a matter of fact, he did have telephones in his house but not one in his study. But perhaps the most important immediate effect of the telephone was to facilitate the burst of urbanization that took place in the 1880's. The rapid growth of modern cities and industry depended on the telephone for instant communication without the delays and congestion of hand-carried messages.

The telephone brought Bell fame and fortune before he was 30. He never ceased inventing, but in his later discoveries he was never able to equal his first achievement. Bell brought together an unusual complex of knowledge, ability, and character in his pursuit of the telephone: speech, music, acoustics, mechanics, and electricity all mixed in his mind and combined with ambition, drive, and an eloquent, compelling personality. His inner need to achieve was strong, and he had pushed himself to make the telephone work in order to marry Mabel Hubbard. With marriage and success Bell's interests and drive were free to follow many paths. Bell next conceived of sending the voice by a light beam and developed a wireless "photophone." He used the element selenium, which varies in its ability to conduct electricity depending on the amount of light falling on it. A transmitter with a voice-vibrated mirror produced a beam of light with a fluctuating intensity. The beam was focused on a selenium receiver which in turn varied the current to a telephone receiver. Bell succeeded in transmitting sound by sunlight but only so far as the beam would go unbroken. The photophone was not practical without a technological breakthrough like today's laser. But Bell's photophone work opened up the way for the nondestructive analysis of opaque substances by means of light rays—photoacoustic spectroscopy —a technique that has only been recently utilized.

As an aid to surgery, Bell developed a telephonic probe which located bullets or other metal in a body. The probe was successfully used until it was supplanted by the X-ray at the time of World War I. Also in the field of medicine, Bell invented the "vacuum jacket," a mechanical device placed around the chest which administered artificial respiration. The concept was later revived in the modern iron lung.

After winning France's prestigious Volta Prize in 1880, Bell organized the Volta Laboratory with two other engineers and inventors. They turned to the phonograph, which Edison had abandoned to begin his electric light experiments. The Volta associates developed a new method to record sound by cutting grooves in waxed cardboard cylinders, as well as a floating (rather than indenting) stylus to reproduce the recording. They successfully marketed the patents, and Bell used his share to establish a trust fund for research on deafness. But it was there that the group's collaboration ended.

His imagination captured by mechanical flight, Bell joined a physicist friend, Samuel Langley, in his efforts to build a flying machine during the 1890's. Bell also experimented with lighter-than-air craft and kites to work out aerodynamic and structural problems. In 1907, inspired by the Wright Brothers' flights, Bell gathered a group of young engineers, including future airplane manufacturer Glenn Curtiss, to form the Aerial Experiment Association. In designing their airplane they devised a means to achieve lateral control using "ailerons" or hinged wing tips which moved in opposite directions. They were later to discover that similar mechanisms had already been developed in France. But the group had only agreed to associate for a year, and although they extended their work for six months to achieve a flight of some three and a half miles at 50 feet, they disbanded in 1909.

While working with kite structures Bell recognized the possibilities of the tetrahedron, a geometric shape of four equilateral triangles forming a pyramid with a base. He envisioned a new kind of architecture, combining both strength and lightness. The advantage of the tetrahedron is that its sides, as triangles, are all compression or tension and cannot be warped. Structures made up of tetrahedrons are braced in three dimensions. The individual members can be mass-produced, stamped out of metal for instance, and then bolted together. Tetrahedral construction, moreover, requires relatively little money, time, and skill, yet it can construct great spans for roofs or bridges. Bell's concept foreshadowed Buckminster Fuller's geodesic domes, and today it has come into wide use as space frame architecture. Bell patented tetrahedral construction in 1904 and built a tower at his Canadian summer home to demonstrate its potential. But the idea did not take, and Bell did not pursue it.

Another offshoot of his aerodynamics research was Bell's interest in hydrofoil boats. With fellow AEA member Casey Baldwin, he built a series of "hydrodromes," the last of which set a world marine speed record in 1919 at 70.86 miles per hour. But neither the U.S. nor the British Navy was interested, and the Bell-Baldwin Hydrodrome Company failed to find any business. The hydrofoil patents granted in 1922 were to be Bell's last.

114

Bell anticipated today's technology in many ways with these inventions, but when compared to the telephone, his later ideas have a sense of anti-climax. Yet Bell's interests were scarcely limited to invention, and it was in these varied pursuits that he found meaning.

If Bell's life was particularly concerned with communication and breaking down the barriers between people, this commitment was nowhere more evident than in his concern for the deaf. In his own family both his mother and his wife Mabel were deaf. Mabel had lost her hearing at the age of five and had come to Bell as a pupil in 1873. They were married four years later. An accomplished speech-reader, she had learned to articulate relatively well if not perfectly. She often astonished new acquaintances with her ability to follow their conversation.

As a teacher of the deaf, Bell's ideas embroiled him in a life-long controversy which still continues among those who work with the deaf: the extent to which sign language as distinct from lip or speechreading should be used. (The form of sign language most widely used in Bell's day had been devised by the Abbé de l'Épée in the 18th century to express ideas rather than just finger spelling.) In general, sign language is less prone to misunderstanding: that is, it does not require the context to fill in the meaning. It is easier, especially for those who are congenitally deaf, and perhaps more natural. Speechreading is much more difficult, relying heavily on quickness of mind to grasp the context and—most importantly —requiring the user to be familiar with the language being spoken. Bell's argument against sign language was that it was a specialized language which cut the deaf off from communication with everyone except those who could use it. He also felt that it deprived the deaf of a full understanding of English—or other language—since sign language falls far short of English in conveying abstract ideas and has a limited subtlety of expression.

As a corollary to speechreading, Bell emphasized teaching the deaf to articulate, to think, and to communicate in English. His beliefs brought him into frequent conflict with Edward Gallaudet, noted educator of the deaf and originator of a "Combined Method," which made extensive use of sign language although employing other methods as well. Bell's commitment also led him to help found and endow the American Association for the Promotion of the Teaching of Speech to the Deaf in 1890. Renamed the

Alexander Graham Bell Association for the Deaf, it still serves as a worldwide information center and library, administering grants and publishing research relating to deafness.

Another of Bell's varied legacies is the *National Geographic Magazine*. Out of a sense of family loyalty, Bell assumed the presidency of the National Geographic Society in 1897 on the death of Gardiner Hubbard, who had founded the organization. To stimulate a declining membership and revitalize the society's dry and pedantic magazine Bell found an able, young editor in Gilbert Grosvenor. Bell's financial contributions and ideas for eye-catching, dynamic illustrations provided the basis for Grosvenor's sweeping innovations in the magazine's format. The new editor created the successful mix of photographs and factual, easy-to-read articles which is the hallmark of the magazine even today. He later married Bell's elder daughter Elsie, and the family is still active in the magazine.

Bell's wide-ranging interests, active participation in many causes and organizations, and gift for public speaking gave the impression of an expansive personality. People constantly sought him out. He seemed almost larger than life with a tall, imposing figure, resonant voice, dark eyes, dark complexion, and full beard—its black growing white as he aged. Yet all his life he struggled against a tendency to withdraw. His chosen hours of work, from late at night through to early morning, reinforced his solitude. He felt a natural sympathy with the isolation of the deaf. More at home in the little world he created for himself on his Nova Scotia summer estate, Bell nonetheless led an active social life in Washington, D.C., where he had settled in 1880. Mabel did much to counter his aloofness as did his own awareness of his reserved disposition.

Alexander Graham Bell sought communication for the world at large, for the deaf in particular, and for himself as an individual. To Americans his name—so appropriate—is inextricably linked with the telephone. Indeed, at the time of his burial all telephone service in the United States was stopped for a minute in tribute. Certainly Bell's achievements place him at the center of the communications revolution that so immediately links people at opposite ends of the globe together.

11

THOMAS ALVA EDISON AND THE RISE OF ELECTRICITY

Thomas Parke Hughes

At the same time that Thomas Alva Edison flourished, the United States emerged upon the world scene as the great technological nation. This simultaneity was not altogether accidental, for Edison drew upon the sustaining environment and, at the same time, helped create it. His most fruitful years were those spent at Menlo Park, New Jersey, from 1876 to 1886, which was about the time America rose to preeminence in invention and industry. By 1890 the United States led the world in the number of patents granted and in its iron and steel production. Furthermore, its production of coal—the basic fuel and an important chemical —ranked second to none. Of the many inventive Americans in this productive era, Edison was the most prolific with no less than 500 patents by 1885 and with hundreds more to follow. Not only did he have the largest number of patents, but the devices and processes they covered were financially rewarding and technologically impressive.

No wonder that Edison was the American hero of the Gilded Age. Newspaper coverage showed him to be one of the most interesting of men to Americans, and a popular poll conducted during his lifetime revealed him as the man that Americans most admired. Americans of his generation liked and admired him because he had risen from humble circumstances to a position of affluence and fame and because he provided them the material needs and pleasures for which many had come to the New World.

Born in 1847 in Milan, Ohio, a Midwestern town earning its living from a canal and from the processing and shipping of agricultural products, Edison was the son of a small manufacturer of wood shingles and of a schoolteacher. Later, the family moved to Port Huron, Michigan, and there, at age 12, he became a traveling candy and newspaper "butcher" (salesman) on the railway to Detroit, setting up his small laboratory for science experiments in the baggage car. This episode and other activities of young Edison became national folklore after his world fame provoked a host of biographies. Parents urged their children to emulate him, and, through him, these parents lived vicariously the American dream.

The anecdotes, once so well known, now recede into the background of the American past as a more worldly America seeks other representative types and even anti-heroes. So we should briefly recall that the laboratory on the car to Detroit had a fire that compelled young Edison to experiment elsewhere. Also we might remember that he was introduced to the telegraph by a grateful operator whose small son Edison had pushed from the path of an oncoming train. And that his partial deafness began to isolate the young man from his immediate environment and turned him toward introspection. To understand him, we should note his *Wanderjahre* as journeyman telegraph operator, for he learned not only the art of survival in a tumultous and demanding world, but a great deal about electricity. He was thought a bit odd by his fellow operators for choosing to stage experiments according to the precepts of the great Michael Faraday, whom Edison much admired and deeply read, instead of joining his fellows in a night on the town, wherever that might have been for the highly mobile telegraphers.

By 1868 he was working for Western Union in Boston, and there he committed himself to invention. At first his inventing could only be during off hours, but he found time to build and patent an automatic vote recorder for which he could find no market. (He later said that it was then he decided to identify a market always before he invented—an obvious strategy employed by most professional inventors, then and now.) No amateur, and determined to

118

live by his new profession, Edison journeyed to New York in 1869. With him he carried ideas for improvements in telegraph systems and a sharp eye and a clear head for opportunity. Opportunities opened to him shortly when chance allowed him to repair a Wall Street printing telegraph at a time when its price quotations were badly needed. Subsequent support from the grateful owner, who was well connected in the telegraph business, gave Edison an entree. Less familiar are accounts of the intricate business and technological activities in which he became involved as he invented telegraph improvements, including a quadruplex design, for several, even competing, telegraph companies. When Western Union fell into the hands of Jay Gould, the notorious financier and waterer of stock, Edison said his kind of inventiveness was no longer needed there, so he became an independent inventor, choosing his own problems, making his own inventions, and forming new companies to market them. In 1870 he had set up a telegraph manufacturing shop and laboratory in Newark, New Jersey; in 1876 when he decided to become an independent inventor, he drew upon his capital and his growing reputation to fulfill a vision—the establishment of an invention enterprise or, as some said, an invention factory.

He chose Menlo Park, a lonely site on the Pennsylvania Railroad between New York City and Philadelphia, which as bases of supply and ready markets were only an hour or so away by train. But at Menlo Park, unlike the cities, there was a freedom from worldly distraction and an invitation to concentration. Edison realized this as he moved old experienced aides such as John Kreusi, the machinist and ingenious model maker, from the Newark shop and brought in new ones who would have to learn the Edison style and absorb the deep commitment to inventing a method of invention.

The compound at Menlo Park was both cozy and workmanlike. The buildings provided the resources needed by a professional inventor, and Edison soon became known as the Wizard of Menlo Park. Within a few years of settling at the new site, Edison had a building for an office and a technical and scientific library (long series of the world's leading journals were housed there for the seekers of ideas about the state of the art), as well as another large building that housed on its two floors a remarkably well-provided chemical laboratory, an electrical testing facility, and, initially, a

Thomas Alva Edison in the laboratory at Menlo Park, New Jersey, exhibits the forerunners of modern radio tubes developed by him and his team of researchers.

machine shop. (Later, the machine shop—the producer of electrical and mechanical models, small and full-scale—was separately housed.) After Edison concentrated upon the invention of a system of electric light, small buildings for blowing the glass bulbs and for obtaining the filament carbon were added. A carpentry shop rounded out the facilities. When the invention factory was built and full of life, a number of watercolors and drawings captured the public imagination by portraying it snow-covered, suggestive of a bountiful Santa Claus and his busy elves. In fact, the place worked more like a center for advanced invention.

The sociology of the Menlo Park group deserves more study. It was far more complex in its interactions than the popular impression of the inventive genius delegating work to eager, pliable assistants. Thousands of laboratory notebooks from Menlo Park suggest that Edison, sensitive to innumerable factors, decided upon the ultimate objectives, but that a handful of men immediately around him engaged in team research and development, coordinated and monitored by him. The historical record also reveals that the members of the inner circle occasionally changed, as did the equipment at Menlo Park, in accord with the nature of the project.

From 1878 to 1882 when activity concentrated upon the invention and development of a system of electric lighting, Francis Upton, John Kreusi, and Charles Batchelor appear frequently in the notebooks carrying out experiments and recording observations. Their particular characters and work tell us much about Edison's. Upton's presence discounts the belief that Edison did not value science. A physicist, Upton had studied at Bowdoin College, Princeton University, and in Berlin under the famous Herman von Helmholtz, before coming to Menlo Park at the start of the electric light project, to read about the state of the art in the foreign scientific and technical literature. Soon he had charge of the development of the dynamo for the system. Charles Batchelor, another intimate, was so deeply involved that some said when he absented himself from the laboratory, Edison suspended work. Batchelor was an Englishman and expert mechanic who had first come to America to work on textile machines. John Kreusi, also a machinist by trade and European by origin, was the master model builder, a function without which Edison and other inventors could not work. It was said that Edison had merely to tell Kreusi of a device to have him reduce it to drawings and a model. (Kreusi is best known for his rendition of the first phonograph.) Edison, as was frequently done by other professionals in technology and industry, drew upon British and European technology and science as he surrounded himself with men who had apprenticed or studied in the Old World.

Edison was the leader of an invention and development group (today it would be labelled research and development), but he was also something more—an inventor-entrepreneur. Only a few men

121

today attempt to carry such a broad range of responsibilities as designated by the term inventor-entrepreneur. Edison not only presided over invention and development, but he also took part in financing, publicizing, and marketing for the project. His most famous project, the electric light system, serves well to illustrate the point.

To choose a problem or a project is a critical decision for an independent inventor; an inventor hired by a corporation usually has guidelines explicitly defined or implicitly revealed by the vested interests of the corporation or agency for which he works. Edison decided, for complex reasons, to expend his resources upon the electric light project in 1878. Friends in science and engineering told him that the state of the art in incandescent lighting suggested that practical achievement might be near. Technical periodicals and patents also signalled activity in incandescent lighting. Such information alerted Edison to the possibility that he might solve the remaining critical problems—such as a durable filament—that would make the difference between ingenious tinkering and commercial success. He had confidence in his ability to solve electric lighting problems because, like so many professional inventors, he knew his characteristics and drew upon the experiences that had helped shape them. In short, he was, after years of work on the telegraph, an expert on electrical matters. The electromagnetic phenomena of the telegraph, the electrochemistry of the battery, the fine mechanics of the relay, and the laws of circuitry, all had relevance to the new endeavor, if one could transfer, adapt, and invent by analogy.

Another reason for working on electric light was the nature of Menlo Park itself, both its physical and personnel resources. For its day Menlo Park represented a considerable investment of resources and, as a result, it had substantial momentum. It had mass, movement, and direction. Therefore, certain problems could be solved best at Menlo Park, and others better elsewhere. Edison and his advisers realized that the problem of inventing and developing an electric lighting system suited Menlo Park, for a system involved electromagnetic machines (generators), delicate apparatus (switches, fixtures, controls, incandescent lamps, and so on), and complex circuitry. Because of the diversity of problems posed by the varied components, the complex of facilities and people at Menlo Park could be advantageously employed. The system required the repeated testing and experimentation for which Menlo Park and its men were also well suited. Most important, a system of electric lighting needed vision, planning, and coordination for which Edison had a genius and for which Menlo Park was designed.

Before finally committing himself, he had to define the market and to identify the financial resources. Edison, who often thought in analogies, saw that an incandescent lighting system was like a gas one, and he knew that gas lighting thrived commercially. Therefore he would be following the less risky course of improving, instead of introducing, a product or service. To make sure, he commissioned a survey of gas lighting use in the thickly populated Wall Street district of New York City. There the deep, man-made canyons between the office buildings and the thousands of offices within needed artificial light. It was not by coincidence that he focused upon offices owned by men such as J.P. Morgan, who had the funds to finance Edison's project—and did.

The project needed a business and financial structure as well as the technological one provided by Menlo Park. So with the advice of Grosvenor P. Lowrey, whose strong characteristics as an experienced business and financial entrepreneur nicely rounded out his own, Edison established in the fall of 1878 the Edison Electric Light Company. Its purpose was to fund the inventive enterprise of Edison insofar as it pertained to electric light and power, and to promote throughout the world the adoption of the patented inventions. There is much to be learned about the art of technological innovation from the constitution of this company. Edison's investment, for which he received stock, was his form of capital—his patents. He assigned to the company for five years all of his patents in the general category of electric lighting and power. J.P. Morgan and other financiers contributed their resources—cash. He awarded his most valued staff members, such as Upton, stock in other companies which were soon established. These companies manufactured components and performed services that would be needed in the Edison system of electric lighting. (Unlike some organizers, Edison did not impose an administrative structure upon the creative concept.) The companies institutionalized the various components, or functions, of the system: the Edison Electric Illuminating Company of New York (1880), founded to preside over the first demonstration central station; the Edison Machine

Works (1881) to manufacture the generators; the Edison Lamp Works (1880); the (Edison) Electric Tube Company (1881) to make the underground conductors for the distribution systems; and Bergmann & Company to produce electric light accessories. Other inventors who came into the lighting field were often willing to modify the designs of existing generators or to turn over manufacture of components to companies not under their direction and control.

The history of the invention of the Edison electric lighting system is well known and need not be repeated in detail here. It should be pointed out, however, that far too much emphasis in most accounts has been placed on the search for—and testing of—the lamp filament. This stress distracts attention from the essence of Edison's genius. He wrote of his approach:

> It was not only necessary that the lamps should give light and the dynamos generate current, but the lamps must be adapted to the current of the dynamos, and the dynamos must be constructed to give the character of current required by the lamps, and likewise all parts of the system must be constructed with reference to all other parts, since, in one sense, all the parts form one machine, and the connections between the parts being electrical instead of mechanical. Like any other machine the failure of one part to cooperate properly with the other part disorganizes the whole and renders it inoperative for the purpose intended.
>
> The problem then that I undertook to solve was stated generally, the production of the multifarious apparatus, methods and devices, each adapted for use with every other, and all forming a comprehensive system . . . .

Rarely does one find so succinctly stated a systematic approach to invention.

By September 1882 the system had been conceived, designed, patented, and tested on a small scale at Menlo Park. The equipment had been manufactured by the various Edison companies. The first Edison central station for public supply was then placed into operation, serving, as planned, New York's Wall Street district with about a one-mile radius. The Pearl Street Station had six steam engines, driving six Edison Jumbo generators, each capable of supplying 1,200 16-candlepower lamps. Within a year, about

8,000 Edison lamps were being supplied from a 110-volt distribution system. The world celebrated its first central station's technical success.

Financial reports show, however, that for the first few years the Pearl Street Station sold electricity at a loss. This was sustained for several reasons. Foremost was the consideration that Pearl Street served as a demonstration plant to interest local civic leaders and financiers throughout the country and abroad in buying the licenses to operate an exclusive franchise and the equipment of a central station similar to Pearl Street. Another reason was the valid assumption that as service was improved, customers were added, unit fixed-costs were lowered, and various economies were achieved through rationalization, the operation would then become profitable. It did, before fire destroyed the historic station in January 1890. By then, there were Edison stations in large cities and small towns throughout the world. The era of the Edison direct-current station had been established; this era in the 1890's gave way to that of the alternating or polyphase station serving a larger area with both power and light over high-voltage systems. Thomas Edison did not demonstrate the flexibility to make the transition.

Edison's period of brilliance passed with the triumph at Pearl Street. He lived and worked on until 1931, adding to this long list of inventions and patents, and making substantial innovations, but his later contributions to the technology of motion pictures, magnetic ore separation, portland cement manufacture, the storage battery, and the derivation of rubber from indigenous American plants lacked the incisive insight and the dramatic rendition of his work on quadruplex telegraph, the telephone transmitter, and the early phonograph, all of which—like the lighting system— came before 1882. Perhaps a key to the apparent diminution of the inventive powers can be seen in the considerable and unsuccessful effort he made to introduce a process of magnetic ore separation.

After success at Pearl Street, other events in Edison's life added up to a watershed—the gradual turn downward after the peaks of achievement. In 1884 his first wife died of scarlet fever, and some of his friends believed that Edison, deeply grieved, then lost his taste for Menlo Park. In 1886 he moved into a new laboratory of his own design at West Orange, New Jersey. It was much larger and more complex than the rural compound so easily pervaded by his personality. Also in 1886 he married Mina Miller, an attractive young socialite from Akron, Ohio, for whom he purchased an estate, Glenmount, in the hills above his new laboratory. (His first wife, Mary, had worked in his Newark shop, and they had lived in a relatively simple house in Menlo Park.) Between 1882 and 1892 he was also losing influence in his electrical manufacturing companies, and at about the time he embarked upon the ore project, he finally sold out his stock in the enterprises and saw them consolidated in 1892 not as Edison General Electric, but as simply General Electric.

Income from the sale of the electrical stock financed the magnetic ore separation project. Edison decided by a market analysis, reminiscent in its thoroughness of that made in the Wall Street district for the Pearl Street Station, that there would be a return on the investment. He concluded that if a low-cost process for concentrating the low-grade magnetite ores of the Northeast could be developed, then, because of lower transportation costs, these could displace in the eastern steel mills the naturally richer ores of the upper Michigan peninsula. He calculated that the concentrated ore could be sold competitively and profitably at the eastern mills for $6.00 to $6.50 per ton. In order to minimize the cost of concentration, he decided to exploit the economies of scale. He invented and developed a massive system of magnetic ore separation. In essence the system broke down large aggregates of rock, sand, and ore into fine particles and then passed them between the poles of a series of powerful electromagnets that drew off the highly magnetic ores and thereby concentrated them. There were innumerable technological problems to be solved, but the process envisioned at the cost projected was in operation after five years in 1899. The catastrophe for Edison was that by then the fabulously rich ores of the Mesabi range in Minnesota were being exploited, and these could be delivered to the eastern mills below his costs. He shortly abandoned the project, having lost at least $2,000,000.

Edison admirers insisted that he had succeeded by technological standards because the process worked as intended. But Edison did not judge himself by these standards; he was an inventor-entrepreneur and, as such, he viewed finance, business, invention, and engineering as inseparably intertwined. Why, then, had he failed as an inventor-entrepreneur? The most obvious answer is that his market projections had not been accurate. Other reasons,

126

however, suggest themselves. In embarking upon the ore separation process, he abandoned familiar ground. Earlier he chose problems demanding brilliant conceptions of elegant devices and machines such as telegraph instruments, telephone apparatus, the electric lamp, the dynamo, and the phonograph. He had not proven himself remarkable as an inventor of large-scale manufacturing systems. Furthermore, there was an excitement about his earlier inventions shared by those who used them and the man who invented them. The design and organization of the ore separation process was a grim and grimy enterprise. For almost five years he and hundreds of workmen labored in the bleak highlands on the New Jersey-Pennsylvania border. One can only ask how much of the Edison spirit and enthusiasm, so clearly displayed in the Menlo Park notebooks, worked for him on the ore project. Also missing were the fine craftsmen and electricians who worked so near him and so well at Menlo Park. No Kreusis, Batchelors, or Uptons emerged as salient figures from the ore separation project. In addition, Edison was losing his way in entering a field where brute size—technological, financial, and business—counted more than ingenuity and finesse. Elmer Sperry, another noted inventor and contemporary of Edison's, remarked that he always chose the hard and fine problems in order to avoid vulgar competition.

Despite the outcome of the ore separation project and despite the lack of brilliance surrounding later projects such as the storage battery, Edison rode the crest of public acclaim until, by the time of his death, he approached the status of a secular saint, a representative to masses of Americans of the best in the American character. They saw him as a plain-spoken, self-educated, practical-minded, eminently successful, native genius. They believed him an inspired empiricist, both an experimenter and tinkerer. He was an American success story—that of a hard-working, hard-nosed, down-to-earth man who provided material in abundance for an upwardly mobile society. The public wanted more of him than inventions; his pronouncements about education, religion, and other general questions commanded front-page newspaper space and became oracular statements for countless admirers.

In fact, Edison was more complex. He spoke plainly when raising money from Wall Street financiers who distrusted long-haired scientists; he was self-educated, but his reading included the classics of Western literature and the notebooks of Michael Faraday; he was eminently successful, but not simply because he had inventive genius; and, truly, he practiced the art of the experimenter with consummate skill, but he also knew the science of his day and used it to formulate hypotheses and organize experimental data. In the laboratory with his associates, he could be described as "the wizard that spat on the floor," but it is unlikely that the floors were so stained at Glenmount. Long before professional public relations perfected image-making, Edison presented to the world an inventor whem it would support and a hero whom it needed.

During his lifetime the center of industrial research moved from Menlo Park and West Orange into the General Electric, Bell Telephone, and DuPont Laboratories. Men with advanced degrees in science working in the laboratories did not simply use the available science to solve their technological problems; they generated the science as needed. Americans, by the time of Edison's death, began to look to men in white coats bending over microscopes as images of the new research and development bringing "better things for better living."

Despite his limitations, however, Edison was nonetheless a representative American and a key to understanding its late 19th-century character, when it became the world's leading technological and industrial power. America, then, had moved beyond steamboats, railroads, and textile mills; it had not yet reached the stage of automatic controls, missiles, and computers. Edison possessed just the touch to provide the lights and sounds that brought a sense of well-being, even affluence, to hard-working people still laboring on the construction site that was young America.

ments stolen from the busy scramble of their highly-competitive picture-taking activities. Even in larger firms such as E. & H.T. Anthony & Co. of New York, only occasionally did personnel pursue an improvement or development. Apparently viewing scientific or technological creativity as a *spontaneous* event that happened quite unpredictably, like being struck by a bolt of lightning, the two Anthony Company owners took as a new partner a Frenchman whose prime responsibility was to search Europe for novelties. In other words, he was to scour the Old World for the creative products of the sporadic lightning bolts as they struck the inspired geniuses.

Given this romantic conception of technical creativity and given the nature of the industry in America, the Anthonys's approach made commercial sense. Most production was limited to small quantities, and it was highly decentralized. Therefore, substantial research investments in the creation of new products would be difficult to recover. Moreover, most American photographers held anti-patent attitudes—attitudes derived in part from the highly controversial patent disputes of Daguerre and Talbot in England. Finally, in a decentralized industry consisting of small firms, strict enforcement of patents was not commercially worthwhile.

In the late 1870's the advent of the much less perishable dry gelatin emulsion created a veritable revolution in production. With it, the character of research changed. Developed in England, factory-made gelatin dry plates were introduced in the United States in the late 1870's. Although the production facilities were initially quite small, they soon grew into large factories with specialized personnel. The most important of these specialized areas was the emulsion-making, where improvements protected as trade secrets soon became critical to the commercial success of a company. Because of the specialization, the emulsion-makers could devote substantial time to improving their art, and, accordingly, some vigorously sought improvements and innovations in photographic materials and their manufacture. With the shift in production of the highly difficult-to-produce photosensitized materials from the photographer to a specialized manufacturer, interest in photography grew among serious, enthusiastic amateurs. Consequently, the market for materials grew, and the financial rewards of innovations increased. Improvements protected by patents and trade secrets played an increasingly important role.

131

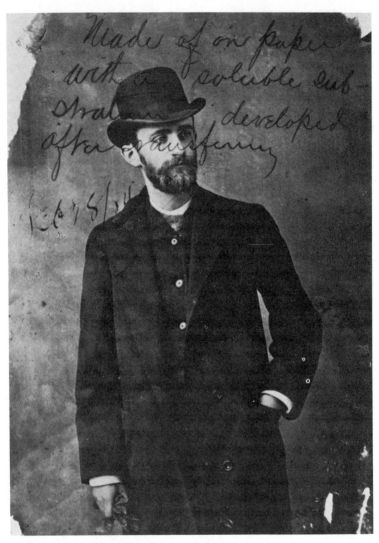

George Eastman made this self-portrait in 1884, noting "Made on paper with a soluble substratum developed after transferring."

Source: Eastman Kodak Company

In such an atmosphere another major revolution occurred in photography. A diminutive, socially-shy yet confident businessman, George Eastman, had entered the industry about 1880 as a dry plate manufacturer with the goal of obtaining a large share of the American market by employing patented machinery. By 1883 he observed that his machinery patent did not effectively limit competition, and, as a consequence, price competition was sharply cutting his profits. Therefore, he hired a local Rochester camera maker, William H. Walker, to join with him in replacing the dry plate by creating a new system—roll film photography.

During the first half of 1884 they created a roll film holder for replacement in the plate holder on the back of the camera; they developed a paper roll film; and they designed special machinery for the continuous production of the photosensitized paper film. The concept of the roll holder was not new. Leon Warnerke had introduced such a scheme in England in the 1870's, but he designed his system only for handicraft style of production, he sensitized his film only with the very slow-speed dry collodion then available, and he did not patent his system. The Warnerke scheme decidedly failed. Nearly a decade later in a review of the Walker-Eastman roll-holder mechanism, Warnerke himself raved about the outstanding characteristics of this substantially improved American version. Yet, the roll film proved less satisfactory. The professional photographers, who constituted the overwhelming part of the market, complained of the poor images attained from printing through paper. When Eastman introduced a film which required stripping the gelatin emulsion from the paper backing for printing, the professionals continued to ignore the film system because of the tedious manipulations required.

By 1887 the Eastman company had invested substantially in research and in the development of this roll film system of photography. Each element of the system—roll holder, film, film production machinery—had been patented. As by-products of the system, the company began production of photosensitized paper and, in order to create a further market for the paper, inaugurated developing and printing service. Both of the by-products were a success, but Eastman came to realize the roll film system itself was a commercial failure. Then in 1887 came perhaps the most crucial moments in the history of photography. Eastman, realizing the failure of his system with the professional market, reconceptualized the market, later remarking:

> When we started out with our scheme of film photography,
> we expected that everybody that used glass plates would take
> up films, but we found that the number that did this was
> relatively small and that in order to make a large business we
> would have to reach the general public and create a new class
> of patrons.

Therefore, George Eastman sought to design a simple-to-operate camera for the amateur, drawing upon the company's developing and printing service to provide all the service, including the tedious stripping of the film, for the amateur. He designed the camera—the Kodak camera. He sold it loaded with a 100-exposure roll of film. After the exposures were made, the amateur photographer returned the loaded camera to the factory for the developing and stripping of the film, producing of prints, and reloading of the camera with fresh film. By reconceiving the market and then redesigning the Kodak system accordingly, he essentially created the mass amateur market for photography, revolutionizing photography and with it the photographic industry. By the mid-1890's mass amateur photography had caught on. Many times the company's rapidly expanding facilities were unable to meet the demands of the Kodak camera snapshooters. Eastman had democratized image-making for the world. At the same time the increased scale of his production permitted greater specialization of function and encouraged commitments to specialized research and development activities.

With the new mass scale of production and marketing, including substantial overseas operations, a new phase of business strategies and conceptions of technological innovation emerged. Where during the 1850's through the 1870's the leaders of the American photographic industry had sought new developments wherever they occurred spontaneously, during the 1880's a few researchers, with faith that some new solution or alternative could be found, had sought specific developments and, by patenting every element of their new developments, had sought legal monopolies. The experience of multinational enterprises operating within the legal frameworks at home and abroad demonstrated that such strategies did not fully protect the products of their research and development investment. At home, patent litigation was com-

The No. 1 Kodak camera (1888), disassembled, was priced at $25 and loaded with enough film for 100 exposures. After exposure, both camera and film were sent to Rochester, the film developed and printed and new film inserted—all for $10.

Source: Eastman Kodak Company

An advertisement featuring an Eastman slogan—"You Push the Button, We Do the Rest"—popularized the No. 1 camera.

Source: Eastman Kodak Company

plex and costly and sometimes required years to come to a decision. By that time the patented feature might be out of date.

Abroad, the patenting of products in a large number of countries was complex, and full protection could not be insured. Likewise, litigation required extended periods of time during which full commercial exploitation might be curtailed. Hence, at Eastman Kodak there emerged during the mid-1890's the strategy of continuous technological innovation. George Eastman succinctly summarized the motivations behind the establishment of an experimental laboratory:

> I have come to think that the maintenance of a lead in the apparatus trade will depend greatly upon a rapid succession of changes and improvements, and with that aim in view, I propose to organize the Experimental Department in the Camera Works and raise it to a high degree of efficiency. If we can get out improved goods every year nobody will be able to get out original goods the same as we do.

Here Eastman clearly states a strategy of annual model change, a strategy dependent upon a steady stream of improvements from an experimental laboratory. On a less formal basis, a few selected persons at Eastman Kodak pursued similar research and development in the photosensitized products departments as well.

Prior to the formal founding in 1912 of the industrial research laboratory at Eastman Kodak, the research and development facilities could boast substantial achievement. This included thin, rolled celluloid film; cinematographic film; continuous methods of film production, including coating and drying; acetate safety film; increased degree and range of spectral photosensitivity of emulsions; considerable color experimentation; plus a host of major and minor improvements in amateur cameras. Hence, industrial research and development was already an ongoing and increasingly important activity in the photographic industry. Yet in 1912 George Eastman took a bold new step with the establishment of the formalized industrial research laboratory. This new institution reflected, in large part, the attitudes and ideas of two men: Eastman himself and C.E. Kenneth Mees.

The immediate stimulus for the establishment of the laboratory came while Eastman was touring Europe in late 1911 and early January 1912. He traveled in France, Italy, and Germany, signing

Kodak's first folding camera, the No. 4, was introduced in 1890.
Source: Eastman Kodak Company

Moving pictures became a reality when Edison (right) used Eastman's flexible film
in his camera. This photograph marking their joint achievements was taken in 1928.
Source: Eastman Kodak Company

negotiated film contracts with major film producing companies. In Germany he accepted the invitation of the executives of the Bayer Chemical Company to visit their plants at Elberfeldt. After touring the plants, he was feted at a luncheon. During the repast he sat next to one of the executives who extolled the company's wonderful facilities, including its research laboratory. He boasted of its equipment and numbers of its research scientists. Then he turned to Eastman and asked, "And how many people do *you* have in your research work?" Eastman apparently embarrassed had to admit that his company had no formalized research laboratory. He returned shortly to England, determined to rectify this embarrassing situation. At once he embarked upon building such a laboratory at Rochester.

Determined to obtain the most promising photographic scientist as the director and architect of the new laboratory, Eastman received from his major European technical-legal adviser the suggestion of young C.E. Kenneth Mees. Eastman approached the tall, gangly Mees, who had just a few years before graduated with a D.Sc. from the Laboratory of William Ramsay at the University of London. Mees was a partner in a small dry plate works at Croyden, just south of London. He was one of the world's leading authorities on color photography and photometrics. Eastman invited Mees to Rochester to establish the laboratory. Mees, concerned about the plight of his elder partner, expressed his reluctance to abandon his partner and the fledgling firm. Determined to have Mees as director, Eastman bought the dry plate works and employed Mees's partner in the company's Harrow works.

Mees embraced the Rochester opportunity enthusiastically and within the year had the new research facility in operation. Eastman, now entering upon a period of substantial philanthropic activity, charged Mees and the laboratory with the responsibility for the "future of photography." More significantly, he committed himself to the support of the laboratory for ten years without expecting any immediate commercial return. Eastman could confidently extend this generous commitment to basic research, knowing full well that Mees was as keenly attuned to commercial results as to basic research.

Mees brought to the laboratory a view of science as an enterprise devoted to accumulating a large body of experimentally determined facts, an ongoing activity which gradually but inevitably progresses ever closer to the true understanding of the nature of the universe. Having abandoned the Fabian socialism of his student days and absorbed William Ramsay's enthusiasm for the application of science to industry, Mees embraced a Baconian perspective that saw scientific knowledge as contributing to society. As Mees then saw it, instead of redistributing the "pie" as the socialists advocate, the enlarging of the "pie" through industrial research was the ultimate solution to the plight of the poor. Accordingly, he saw the prototype of the industrial research laboratory not just in Whitney's General Electric laboratory or Rintoul's laboratory at Nobel, but in Francis Bacon's "House of Solomon," which he described as follows:

> A great number of observed facts would be collected, and from them the fundamental processes of nature could be understood. In this way, he believed, it was possible to attain to "the knowledge of Causes and secret motions of things, and the enlarging bounds of Human Empire, to the effecting of all things possible." This was a great vision, a new vision on earth, and a vision that has in some measure been realized. The method that Bacon suggested for carrying out this idea was the organization of a research institute, which he entitled the "House of Solomon" and described in his "New Atlantis." This institute was to be manned by a great company of Fellows to whom Bacon . . . allotted specific functions . . . . A noble dream, much before its time and greatly overorganized, it led to the idea of cooperation in the pursuit of knowledge.

Eschewing the dependence of research laboratories upon scientists of exceptional caliber, Mees held:

> . . . that much scientific research depends upon the accumulation of facts and measurements, an accumulation requiring many years of patient labor by numbers of investigators, but not demanding any special originality on the part of the individual worker . . . . He can make valuable contributions to scientific research even though he be entirely untouched by anything that might be considered as the fire of genius.

Furthermore, Mees sounded the death knell of the earlier heroic view of invention and discovery:

> It is, indeed, a matter of doubt how many of the men
> commonly considered to be of great genius by virtue of some
> important discovery they have made really possessed any
> distinguishing ability compared with their fellows who did not
> have the fortune to make a similarly important discovery.

Mees organized his laboratory accordingly and established the weekly conference system and various bibliographical tools in order to enhance the cooperation among his staff. The multistoried building housed a large library, several laboratories, photographic galleries, and a small-scale production facility. In 1913 he had a staff of 20 and operated on a little over $50,000; by 1920 the staff had increased to 88 and expenditures to nearly $350,000.

During the first decade and a half the laboratory contributed new understandings in photochemistry and in photometrics. Moreover, it introduced small-scale production of fine organic chemicals when the United States was deprived of its usual German sources during World War I, a home movie system with safety film, and the first color movies. In addition it sharply increased the range and degree of sensitivity of the company's emulsions through the discovery of the catalytic effects of sulphur compounds in silver halide emulsions. Truly Eastman found the laboratory not only fulfilling its promise of being responsible for the "future of photography" but also for the "fortunes of Eastman Kodak."

During his industrial career George Eastman did more than create mass amateur photography and thereby democratize image-making for the world. He embodied those ideas and values which—along with many others in American and European industry—encouraged him to turn from the romantic view of the spontaneous and sporadic origins of new technological ideas and gradually adopt the view that prompted continuous pursuit of technological innovation. The research laboratory, based upon cumulative, anti-heroic views of the origins of new science and technology, stood saliently as the institutionalization of that continuous pursuit.

At the same time the emergence of continuous innovation was

also a business response to the legal complications at home and abroad, further fostered by the growing availability of academically trained chemists, physicists, and engineers. Hence continuous innovation was a culturally shaped response to a certain set of broad social conditions. But, equally important, it reflected certain social values inherent in Western culture, specifically, the high value given to new and improved possessions. In American culture, for example, despite its egalitarianism, society had set aside its anti-monopoly attitudes in the development of new technology. The Constitution provided for a patent system and by law the inventor was granted a legal monopoly for a period of 17 years. Moreover, late in the 19th century Americans lionized their patent-seeking inventors such as Edison. At the same time, they concretized their anti-monopoly attitudes with the passage of the Sherman Anti-Trust Act of 1890.

Thus the forces behind the development of industrial research are multifold and complex. To understand the founding and evolution of this research activity, which is of such importance today, it is necessary both to consider the personal motivations and the ideas of science and engineering and at the same time to study the economic, social, and institutional structures which in turn reflect the values of this society. One of the steadfast values of the past 200 years in America is that technical change is good. The emergence of industrial research laboratories was one manifestation of that value as it filtered through a complex set of cultural and social structures.

13

ELLEN SWALLOW RICHARDS:
TECHNOLOGY AND WOMEN

Ruth Schwartz Cowan

Ellen Swallow Richards was an extraordinary woman whose life had a profound impact upon the history of American technology and upon the ways in which Americans lead their daily lives—yet she did not exert her influence in the usual way. She did not invent a new device nor find a way to manufacture a device that someone else had invented. Rather she changed American technology by the force of her ideas and by the educational movement that was founded on the basis of her ideas—the home economics movement.

Ellen Swallow was born in 1842 in a small New England village. Her parents were "middling people," schoolteachers, occasional farmers, and owners of the village store. Ellen was their only child, and they educated her at home. In her teens she became a schoolteacher herself but did not find the work fulfilling; she had a deep desire for wider acquaintance with the world and for a more extensive education. Consequently in 1868, at the age of 25, she entered Vassar College in Poughkeepsie, New York, one of the very few institutions for higher education for women in the United States at that time. Most of the students at Vassar were younger than she was and considerably wealthier; she supported herself by tutoring them.

Upon graduating in 1870 she resolved that she wished to do something scientific and practical with her life and applied to become a student at the Massachusetts Institute of Technology. MIT had been founded just nine years earlier and had never before admitted a female student. The authorities made a special arrangement which must have been ironically pleasing to Ellen Swallow—they allowed her to attend classes but refused to let her pay tuition so that she would not be a student officially. Official or not, she was awarded the degree of Bachelor of Science in 1873, the first woman to be accorded that distinction at MIT. She remained at MIT for additional study for two more years, serving as assistant to the professor of chemistry, William R. Nichols, who was then engaged in analyses of public water supplies for the Board of Health of the City of Boston. During her postgraduate years Ellen Swallow met and married Robert Hallowell Richards, professor of mining engineering at MIT.

In the ensuing years her combined interest in science and in the problems of women did not flag. Through her marriage she came in contact with some of the wealthier members of Boston society, and she managed to convince some of her new friends to help her in various endeavors. One of these was the Women's Laboratory on the grounds of, but not officially connected with, MIT. Here Ellen

The Women's Laboratory at the Massachusetts Institute of Technology was headed by Ellen Swallow Richards as a separate facility until women were admitted as regular students at MIT in 1882.

Source: Massachusetts Institute of Technology Historical Collections

Richards volunteered to teach the rudiments of science to young women who wished to become schoolteachers. Another favorite project was the Society to Encourage Studies at Home, a series of correspondence courses intended to help homebound women acquire an education. Here Ellen Richards was the author of the science course. Yet another was the New England Kitchen, where the diets of working-class families were studied and public demonstrations were given to show the women of those families how to prepare less costly and more nutritious meals. Here Ellen Richards functioned as chief nutritionist and occasional lecturer. On her own she was also working as a consultant in industrial chemistry; utilizing her training with Nichols she analyzed, among other things, the arsenic content of wallpapers and various adulterants in foods.

In 1882, after considerable prodding from Richards and her friends, MIT decided to admit women as regular students (thus closing the Women's Laboratory), and two years later she was appointed as its first female instructor—in sanitary chemistry—and assistant in the laboratory for the chemical study of sanitation (under the direction of her earlier sponsor, William Nichols). From that time until her death in 1911 she regularly taught courses in the analysis of air, water, and sewage and routinely participated in the research projects of the laboratory. She was thus not only one of the founders of the discipline of sanitary engineering in the United States but also among the first women in the United States to be formally employed as a scientist. Those accomplishments mark her as one of the most unusual women of her time, but the fact that she did all those things while maintaining a stable marriage makes her even more unusual. Among the very few women who achieved professional status in the 19th century there were even fewer who were married. Ellen Swallow Richards was fortunate in being married to a man who encouraged and assisted her in her scientific pursuits. She was also fortunate (at least in terms of her career) in having been childless, as the problems of rearing children while maintaining a career were well nigh insurmountable for most of the women who attempted it in the 19th century.

Despite her contributions to the development of sanitation engineering, Ellen Swallow Richards's influence on American technology really lies in another direction. Roughly at the same time that she began her scientific career, educational reformers throughout the United States were beginning to worry about the need to educate young women in the craft of housekeeping. In earlier times this need had scarcely existed as most girls learned housekeeping either from their mothers or from their employers, but by the middle years of the 19th century this traditional pattern of education had begun to collapse. America was industrializing. It was opening its eastern doors to immigrants from Europe. Girls who would have worked as house servants a generation earlier were now working in factories. Other girls who would have learned housewifery from their mothers now had mothers who did no housework at all but who relied on immigrant servants and new supplies of industrial goods. When these girls became housewives in their own right, they often did not have access to the same luxuries, but the schools in which they had been educated eschewed practical training as "undignified." For other young women the conditions of their housewifery were so unlike those of their mothers' as to make their mothers' teachings useless—consider, for a moment, the situation of a young woman who was reared in a city but found herself setting up housekeeping on the frontier, or that of a young woman who was reared in a peasant household in Europe who found herself setting up housekeeping in a tenement in New York. All of these young women needed training in housewifery, and the educational reformers began to campaign for the inclusion of domestic science (or domestic economy) in school curricula.

The reformers must have been responding to a felt need because their efforts met with success rather quickly. In the West domestic science courses were frequently offered at the agricultural and mechanical colleges that were created after the passage of the Morrill Act; Iowa State Agricultural College seems to have been the first, with its courses in housekeeping in 1872 and in cooking in 1877. In the East the pattern was somewhat different because the women's colleges then in existence were not hospitable to practical subjects. In cities such as Boston, New York, and Philadelphia instruction was offered in "cooking schools" especially created for the purpose. Many of these schools were intended for two different types of students: one set of courses was offered for young, fairly well-educated women who needed instruction in the fine points of running a comfortable home; another set of courses was directed to the poor, uneducated girls who wanted to get good jobs as servants or who were ambitious enough to want to learn the American way of housekeeping. The early proponents of domestic science also discovered that they could easily spread their ideas through the written word; such works as Catherine Beecher's *The American Woman's Home* (1869) and Fannie Merritt Farmer's *The Boston Cooking School Cook Book* (1896) went through many editions in very few years, spreading the gospel of household engineering far abroad in the land.

Early in her career Ellen Richards became a convert to domestic science; its principles, such as they were, pervaded her private life as well as her professional activities. She tried to organize her own home on efficient and healthful lines—eliminating heavy carpets (which collected too much dust), installing many new devices (a gas stove, a vacuum cleaner, hot water heaters, ventilators) as soon as they came on the market, cooking very simple meals while trying to insure that they were nutritious in every respect. Despite the unusual pattern of her own life, Ellen Richards's views about household and family were conservative in many respects. She believed that the home was the center of civilized society and that educated women, if properly trained, would be the best homemakers of all; they could understand (as untrained women could not) the ways to make their homes comfortable, efficient, healthful and "democratic" (i.e. servantless), while avoiding drudgery and leaving themselves sufficient time for other pursuits:

> The educated woman longs for a career, for an opportunity to influence the world. Just now the greatest field offered to her is the elevation of the home into its place in American life . . . . Who is to have the knowledge and wisdom and time to carry out the ideals and keep the family up to these standards? Who indeed, but the woman, the mistress of the home, the one who chooses the household as her profession . . . because she believes in the home as the means of educating and perfecting the ideal human being, the flower of the race for which we are all existing; because she believes that it is worth while to give her energy and skill to the service of her country and age.

During the 1880's and 1890's the goal of training the educated housewife in the precepts of domestic science became the focus of many of Ellen Richards's professional activities outside of MIT.

Ellen Swallow Richards

Source: Massachusetts Institute of Technology Historical Collections

She campaigned to have domestic science included in the curriculum of the Boston high schools, helped to organize the first department of domestic science in an institution of higher learning in the East (Simmons College in Boston), organized the Rumford Kitchen demonstration at the World's Columbian Exposition in Chicago in 1893, and became a leader of the National Household Economics Association which was founded after the Exposition. This association attempted to organize women on a volunteer basis to carry out the kinds of programs that Ellen Richards had pioneered in Boston.

The fact that the women who worked under the aegis of the National Household Economics Association were volunteers is particularly significant because it points up Ellen Richards's other contribution to the development of the home economics movement. She believed that domestic science should be promoted as a suitable career choice for women who were well educated (particularly in the sciences) but who would not, for one reason or another, become homemakers and mothers. Indeed she thought that teaching or practising domestic science would be an ideal career choice for such women because it would not violate the norms for female behavior nor would it threaten the professions that women were barred from entering. Ellen Richards worked to create courses in domestic science because she wanted some women to be able to learn the craft and other women to be able to teach it; she wanted to create a new profession for women.

To this end she invited a dozen women who had pioneered careers in domestic science to meet with her at a summer resort in Lake Placid, New York, in September 1897, to discuss what might be done to encourage the professionalization of their field. This meeting was so successful that it was repeated for ten additional summers until, in August 1908, the group, by now considerably enlarged, announced that it was forming itself into the American Home Economics Association. This association continues to this day to serve as the professional nexus for home economists; in its time it served to formalize the new profession and to give it a name. (The earlier volunteer group, the National Household Economics Association, had merged with the Federation of Women's Clubs several years earlier.)

Within very few years the membership of the association had grown considerably. By 1930 it was possible for young women (and some men) to take undergraduate and advanced degrees in home economics in most of the major universities in the United States, and it was also possible for them to find jobs in the field.

During the early decades of the 20th century home economists came to be employed by school districts, government agencies, social welfare institutions, and a host of business organizations —to teach, to plan diets, to promote goods, to provide homemaking services, to do research.

In the minds of Ellen Richards and her colleagues the goal of the home economist was to bring the individual household out of the Middle Ages and into the 20th century by integrating it with the technological world of which it was a part. Women, they thought, could not possibly become valued members of an industrialized society if they continued to do their daily work in an unscientific, unmechanical, inefficient, and irrational fashion.

> The work of homemaking in this scientific age must be worked out on engineering principles and with the cooperation of trained men and trained women. The mechanical setting of life becomes an important factor, and this new impulse is showing itself so clearly today for the modified construction and operation of the family home is the final crown or seal of the conquest of the last stronghold of conservatism, the home-keeper. Tomorrow, if not today, the woman who is to be really mistress of her house must be an engineer, so far as to be able to understand the use of machines.

This goal, which Ellen Richards enunciated in 1910, was achieved remarkably quickly, in part because American manufacturers began to supply American households with the goods and services that were needed, but also in part because the profession of home economics formed exceedingly quickly and helped to guarantee acceptance of the new products throughout the land. As teachers, the home economists trained their pupils to become consumers; as dieticians, they became major consumers of the new goods themselves; as employees of businesses, they advised on what would be acceptable to homemakers and helped to promote the goods they had advised about; as advisors and helpers of the poor, they strove to assist these people to achieve the status and the behavior of middle-class consumers; as scholars and researchers, they concerned themselves with ways to root out and destroy sources of inefficiency and economic irrationality in households. Through all this, very few of them seemed to notice that their careers were ironic in much the same way that Ellen Richards's had been; part-time housekeepers themselves, they made their livings by trying to convince other women that full-time housework was (or could be) the most rewarding career of all.

This aside, it is nonetheless true that Ellen Richards's ideas, as they were propounded by her disciples, had a mighty impact on American homes in the 20th century. Before the turn of the century the vast bulk of the work in those homes was done by human muscle power, much of it female. In most homes, rich or poor, water was hauled, laundry was scrubbed, furniture was dusted, fires were maintained, food was preserved, gardens were tended, and meals were prepared almost entirely by hand. Within decades much of that work had either been taken over by machines or eliminated or profoundly reorganized. By 1930, 85 percent of American homes had been electrified; by 1940 half of the population enjoyed the benefits of some kind of washing machine; by 1950 almost every home had indoor plumbing and hot and cold running water; and by 1960 there was almost no one left who cooked on a wood or coal-burning stove. Housework is still time consuming, and for some women it is still onerous, but it is not onerous in the ways that it used to be. The increasing freedom that American women enjoy from household drudgery is part of the legacy of the home economics movement as is their increasing willingness to accept and to pay for machinery in their homes. If Americans as a whole have a love affair with machines, this is at least in part due to the fact that most of them live in intimate contact with machinery from the time that they are small children—and this is also part of the legacy of the home economics movement. The ideas that fired that movement, and indeed the very existence of the movement itself, were part of the legacy of Ellen Swallow Richards.

14

GIFFORD PINCHOT AND THE AMERICAN CONSERVATION MOVEMENT

Samuel P. Hays

During the three decades between 1890 and 1920 a new spirit of concern arose for the future of natural resources in the United States—a spirit which has been called the "conservation movement." Contrary to the spirit of the more recent environmental movement, it did not involve prolonged belief that the nation was running out of resources or that there were fundamental "limits" to the earth. Rather it constituted both a reaction against the hitherto wasteful use of resources and the assertion that, with new scientific knowledge and new techniques, resource waste could be ended and even new heights of material production be reached. This phenomenon of the "Progressive Era" called forth significant expressions of the "scientific management" movement; its advocates believed that through systematic organization and centralized direction of resource management, the wasteful practices of the past could be turned into the "efficiency" which would lead to a bountiful future.

Gifford Pinchot, the first Chief Forester of the United States, was one of the architects of this movement, and forest management came to be thought of as one of the central features of the "conservation movement." That movement was, to be sure, far more extensive than forestry alone, for it involved the efficient development and management of the nation's waters as well, and a host of collateral resource problems. But Pinchot, dramatic and flambuoyant in personality, close to President Theodore Roosevelt and active in partisan politics, often played a role on center stage. Historians have continued to associate the early 20th-century conservation movement with him only second to the personality of the President himself. As Chief Forester, Pinchot dramatized the development of the U.S. Forest Service as central to conservation objectives. It is fitting, therefore, that the subject of the role of science and technology in the early 20th-century resource conservation movement should focus particularly on Gifford Pinchot.

Pinchot was the first professionally trained forester in the United States. After graduating from Yale in 1889, he studied forestry in Europe. He came onto the American forestry scene in the 1890's, first as manager of Biltmore, the estate of the Vanderbilt family in North Carolina, where he experimented with "scientific forestry" practices, and then as advisor to federal officials. Scientific forestry programs arose first in the U.S. Department of Agriculture, under which the Bureau of Forestry was established with Pinchot as its head in 1898. The national "forest reserves," as they were then called, however, were under the Department of the Interior, which had long supervised the public lands. Through Pinchot's vigorous political action these "reserves" were transferred in 1905 to the Department of Agriculture, renamed "national forests," and placed under Pinchot's bureau, which soon was renamed the "United States Forest Service" with Pinchot as Chief Forester. He served in that position only five years, but in that short period of time he established the outlines of national forest policy for decades to come.

A large measure of Pinchot's influence came from his role in establishing forestry as a profession in the United States. He inspired a considerable number of young people to become professional foresters. He was instrumental in founding the Yale School of Forestry, where for many years he was a professor. Throughout the nation many of his protégés took positions in forestry schools attached to the agricultural colleges and universities, and in state bureaus and divisions of forestry as well. He played a major role in the development of forestry journals and magazines, in the conduct of associations such as the American Forestry Association, and in the development of professional societies such as the Society of American Foresters. All this, it should be noted, did not come without considerable friction, for Pinchot was an irascible individual who did not suffer willingly those with whom he disagreed. His life, in fact, was strewn with strained friendships as well as intense loyalties. He remained the commanding figure in professional forestry for many years while serving as forestry commissioner and later governor of the state of Pennsylvania. Even after his death in 1946 his memory was invoked more than once in debates over forest policy.

Pinchot appeared on the scene at a time when increasing concern was being expressed as to the future of timber resources in the United States. In the northern and eastern parts of the nation considerable cutting had occurred during the last several decades of the 19th century, resulting in devastated wastelands, repeated forest fires, and abandonment of cut-over lands. All this had made an intense impression upon many Americans. A variety of activities had testified to a concern for future timber supplies, among them the establishment of the American Forestry Association in 1875. The problem was a simple one: Were not future supplies of wood and timber endangered by such wasteful practices? How could one conduct harvest activities with a greater eye to long-run supply? It was in this context that decisions were made to establish "forest reserves" from the existing national public lands in the West. These would constitute the nation's future timber supply.

To Pinchot mere forest reserves were not enough. "Scientific forest management" in the form of sustained-yield practices should be applied. This was the rather straightforward concept that trees for wood production should be considered as a crop and should be cut each year only to the extent of the annual growth. A stand of timber could be described in terms of its total annual growth calculated in terms of board feet. Annual cut would then be limited to this amount, thus guaranteeing a continuous or sustained yield of wood—hence the term "sustained-yield forestry." There were many other aspects of scientific forestry as well. There was the need to regenerate forests and especially the recently cut-over areas; forestry programs required the production of seedlings in forest nurseries and then planting in the field. Fire protection was required. The felled trees left by loggers in the woods gave rise to

153

spectacular forest fires—in the Great Lake states often wiping out whole villages—and created stark recognition of the need to control this devastation. Fire-prevention and fire-suppression programs were critical elements of modern forestry. And there was the need for timber-stand improvement practices, for thinning the growing forest in order to produce a healthy growing stock for future harvest. To these could be added the need for inventories of standing timber upon which sound management practices could be based as well as research ranging from investigations into wood utilization, to "forest influence" (for example, on water supply), to insect and pest control. All this and more came to constitute the elements of "scientific forestry."

Let us go back a moment and place this movement for scientific forestry in a wider historical perspective. To some extent the context is one of a change in the attitude of Americans toward their abundant natural resources. For almost three centuries, from the initial settlements on the Atlantic Coast in the early 17th century through the 19th, Americans had thought of their nation's resources as inexhaustible. If some were exploited, there were more to come. The timberlands of the East were cut down to be replaced by new supplies from the Great Lake states; when these began to run out, new forests were felled in the southern Mississippi Valley; and after these came the Douglas fir forests of the Pacific Northwest. And the same for farm lands, for minerals, for water, and for energy. Charcoal from wood was followed by coal, and coal by oil, each found in greater abundance to the west. The productive wheat farming of the Delaware Valley gave way to that of western New York state, then to the eastern Midwest and finally to the "grain belt" states of the Dakotas, Nebraska, and Kansas. The technologies of the 19th century—the new machines in farming, logging, and mining—only served to quicken the pace of resource exploitation. Substituting machines for human labor, they permitted resource extraction that was increasingly more rapid and more extensive.

In the late 19th and early 20th centuries, however, these "extensive technologies" began to be replaced by more "intensive technologies," which emphasized more systematic management and "efficiency." Science and technology came to stress themes far beyond the mere substitution of machines and mechanical power for human labor, emphasizing those of "efficiency," planning, coordination, and the elimination of waste. The application of technology to resources began to stress long-range planning for continuous production, the use of by-products once discarded as uneconomical to utilize, the application of principles of "scientific management" to the organization of resource extraction, processing, and marketing. At times these new developments stressed the need to check the "wasteful exploitation" of past resource use; but more frequently they emphasized the possibilities for the future which could be created by applied science. They occurred, in fact, amid a set of expanding horizons as to the possibilities of human improvement, of "progress" that could occur by means of the application of systematic empirical inquiry and management control to resource problems.

It was this spirit that came to be embodied in the "conservation movement," of which "scientific forestry" was an integral element. But it was not confined to forestry alone. Equally important was the new perspective that arose from the application of large-scale engineering works to the nation's rivers: the harnessing of water, long flowing unused to the ocean, to "useful" purposes for society. The conservation of water by the construction of reservoirs to be used later for irrigation, water supply, navigation, hydroelectric power, and the reduction of flood damage, came to be a major theme of the entire conservation movement. Originating in the U.S. Geological Survey, later appearing in the U.S. Bureau of Reclamation, and culminating finally in the 1930's in the Tennessee Valley Authority, multiple-use river development was a major expression of the new "scientific management" approach to resources. As time went on, in the 1930's, a similar approach was applied to the problem of soil erosion as techniques were developed to prevent the depletion of this major natural resource—the "capital" upon which agriculture depended—and to establish the scientific management of soil. All these efforts were described by the term "conservation," and those pertaining to water development and forest management were particularly characteristic of early 20th-century conservation.

This movement was shaped heavily by a new breed of scientists and engineers, who were applying even more intensive amounts of empirical inquiry, engineering works, and management techniques to the development of natural resources. The core of this group consisted of officials in the Theodore Roosevelt administration

155

between 1901 and 1909 and included such men as Frederick Newell of the U.S. Bureau of Reclamation, Secretary of the Interior James R. Garfield, Pinchot, and scientist-philosopher W.J. McGee. McGee particularly gave a heavily philosophical rationale to the movement, stressing classic utilitarian themes of the use of science for the "greatest good for the greatest number" of human-kind. Such leaders thought of themselves as substituting "scien-tific" for "political" approaches to resource management, of bringing modern system and order to previously chaotic and wasteful practices. Centralized "efficient" management under the guidance of science and technology constituted the spirit of the movement which they led.

The "efficient development" theme of the conservation move-ment was an intergral part of Gifford Pinchot's concept of scien-tific forestry, and it is important to emphasize this particular twist which he brought to the management of the National Forests. There were at the time competing ideas as to what forest lands should be used for. In fact, the initial forest reserves in the West were established not so much for protecting future timber re-sources as for protecting watersheds for urban water supply. In this context, one of the most controversial issues in forest man-

Gifford Pinchot, as Chief Forester, boards a light plane to inspect the Sequoia National Forest in California.

Source: National Archives

agement at the time when Pinchot became Chief Forester was the impact of sheep and cattle grazing on western watersheds. Advocates of the early reserves established in the 1890's fully intended grazing to be excluded from those lands because of their detrimental effect on water supply through erosion and sedimentation. By 1900 the adverse reaction of cattle and sheep grazers to such pressures to ban grazing, in fact, endangered the very existence of the forest reserves.

From the very beginning of his responsibility for the forest reserves Pinchot took a stance emphasizing the "economic use" of the forests. It was for this reason that he brought about the change in name from "forest reserve" to "national forest." He chose to emphasize the direct economic benefits of forest lands and their direct commodity use. He brought into the national forest administration a leader of the western cattle industry to develop a program of grazing in the forests. This innovation, not explicitly provided for in law but developed by administrative action, Pinchot felt, was essential in order to provide political support for the national forest program. But it also was more in keeping with his basic orientation of looking upon the forests not in terms of "protection" but in terms of commodity production.

Pinchot's stress on the "economic use" of the national forests had much broader implications than simply to increase the role of cattle and sheep grazing. For it involved him in prolonged and persistent controversy over public land issues and especially the conflicts between "commodity" and "aesthetic" uses. His commodity use orientation led to persistent differences with those who wished to reserve lands from lumbering because of their aesthetic value as parks. In the early 20th century the movement to establish national and even state parks competed heavily with the national forest movement, and Pinchot took his stand for the latter and against the former. The national park movement, which had originated with the establishment of Yellowstone National Park in 1872 and which by the early 20th century had resulted in the creation of a half dozen such parks, involved a management philosophy that standing trees should not be cut but should be allowed to mature and decay as a natural system. To Pinchot, and to the overwhelming majority of professional foresters of the day, this was a "waste" of resources.

As Pinchot was seeking to secure jurisdiction over the "forest

157

reserves'' by transferring them to the Department of Agriculture, he was also seeking a similar transfer of the national parks from the Department of the Interior to his jurisdiction in order to apply commodity management principles to the parks. But leaders of the national park movement, such as J. Horace McFarland of the American Civic Association and John Muir of the Sierra Club, knew well Pinchot's objectives and bitterly resisted his efforts. In this move Pinchot failed. At the same time, park advocates sought to establish a separate administration for the national parks, which they achieved in 1916 when the National Park Service was established in the Department of the Interior. This objective Pinchot vigorously but unsuccessfully opposed.

These were not minor or idle differences of opinion; they constituted major controversies over the proper use and management of forested lands which emerged sharply in the early 20th century and have continued to the present day. Pinchot set the tone of the forestry profession. He opposed the establishment of individual national parks in the West, a number of which were carved out of national forest lands; he objected to the protection of the Sequoia trees in California; he sought without success to open the largest natural park in the East, the Adirondack State Park in New York, to commodity timber cutting. In developing his position he contrasted the ''conservation'' approach of ''wise use'' of the forests with the ''preservation'' approach of those who emphasized forests as aesthetic resources. Down through the years, the Forest Service has fought a rear-guard action against ''aesthetic use,'' has objected to carving national parks out of national forest land, and has accepted only with continued reluctance a program of designated ''wilderness areas,'' where commodity timber management is not permitted. It is not too much to say that the association of ''wise use'' with commodity use and the equally strong reluctance to accept fully ''aesthetic use''—let alone a ''wise use''—on the part of the Forest Service and the forestry profession at large was established by Pinchot in the early 20th century. A more accommodating attitude might well have led to the national forests and the national parks being administered by the same federal agency.

Pinchot's drive for scientific forestry also stood apart from another growing ''use'' of the forests which increasingly came into conflict with commodity production—outdoor recreation—that is, fishing, hunting, hiking, and camping. After the widespread extension of automobile ownership and use and the development of improved roads in the 1920's, outdoor recreation expanded rapidly in the United States. Much of this took place on public forest lands. And much of it did not fit well with the forestry profession's heavy emphasis on commodity use of the forests. Outdoor recreationists came into conflict with commodity production in a number of ways. There was the initial competition between game animals and livestock for available forage; Pinchot's decision to legitimize grazing for cattle and sheep on the national forests was a deliberate choice to reduce wildlife objectives in forest management. There were the problems of management for camping and its adverse impact on smoothly-run forest protection activities—for instance, careless campers were allegedly responsible for a number of forest fires. And fishermen protested that timber cutting interfered with their sport by reducing water quality and damaging the aesthetic quality of the forest. Most often contention resulted from simply diverting scarce forest management funds and personnel to outdoor recreation when they could have been used for the scientific wood management programs preferred by foresters.

It was not that professional foresters did not enjoy the out-of-doors. Pinchot was an inveterate woodsman. But those in the hunting and fishing fraternity often found that their use of the woods did not jibe with the plans of professional foresters. Foresters, in turn, considered outdoor recreationists to be bothersome rather than important and significant users of the forest. It was not until after World War II that outdoor recreation secured a more "legitimate," fully recognized place in U.S. public forest management, and even then, despite the widespread public use of the forests for recreation, it seemed to take a back seat to more predominant wood production uses.

Interestingly, President Theodore Roosevelt, a big game hunter and outdoorsman in his own right, had continuous contact with many of the outdoor leaders of the time, who were also personal friends, such as George Bird Grinnell, editor of *Field and Stream* magazine. Some historians, in fact, argue that Grinnell was more influential with Roosevelt than Pinchot. That may be. Yet in a number of critical issues, such as the conflict between livestock and wildlife on the national forests, Roosevelt and the big game hunting organization, the Boone and Crockett Club, which he helped to found, were persuaded to keep silent in order to work out a political compromise with the sheep and cattle interests. Outdoor recreation, in fact, became an integral part of forest management and shaped its development far more rapidly on state forest lands than on federal.

One can sense in these remarks some implications of the early scientific forestry movement and conservation for the present day. Most significant is the way in which the term ''conservation'' has given way to the term ''environment.'' And not without reason. For while the conservation movement was concerned primarily with the manner of resource development, with a stress on efficiency and centralized direction, the environmental movement has been concerned primarily with the ''quality of life'' and specifically the ''quality of the environment'' surrounding the home, work, and play of Americans. The change reflects a host of so-called ''post-industrial'' values—those associated with people's non-occupational world, rather than with how they make a living—values dealing with home and recreation more than with job. Such value preferences reflect levels of personal and social income which provide adequately for both necessities and conveniences and permit special focus on amenities.

The environmental age has several implications for our brief analysis of the role of Gifford Pinchot in the early 20th-century forestry and conservation movements. First is the more general observation that the tendencies toward scientific management at the heart of the conservation movement are, in fact, in deep conflict with tendencies toward the amenity values of environmental quality. The theme of efficiency implicit in the conservation movement logically tended toward more intensive and more large-scale development, toward more comprehensive application of scientific management. But now the theme of environmental quality persistently comes into conflict with such tendencies. While multiple-purpose river development, for example, was the epitome of the conservation movement, and the Tennessee Valley Authority its most successful result, the environmental movement emphasizes the amenity values of free-flowing streams and opposes dam construction by both the TVA and the U.S. Army Corps of Engineers. What was once the triumph of the conservation movement has now become a major target of environmental attack.

Similar tendencies characterize modern scientific forestry.

When the major emphasis of forest management was protection, reforestation, and selective cutting, the U.S. Forest Service, despite its controversies with the National Park Service, was held in high esteem. But more recently there has occurred an increased emphasis on "environmental forestry," on the use of the forest for its amenities as an environment for home, work, and recreation. Such a view is in sharp conflict with the extension of scientific forestry in our own day, with its emphasis on increasingly intensive production, on genetic selection of seed, fertilization, row planting and cultivation, the use of pesticides and herbicides, and clearcutting. Controversies between conceptions of the forest as an environment and as a source of commodities are intense; although they are most sharply etched in the disagreements over designating wilderness areas to be left untouched by foresters and other developers, they occur also in the management of the general forest. While the tendencies which Gifford Pinchot set in motion are represented by intensive forest management, those represented by John Muir and the national parks movement are closer to the historic roots of the modern environmental forestry movement.

And what of the role of science and technology in all of this? Those concerned with commodity production are apt to associate science and technology exclusively with their concerns and to dismiss environmental quality as a rejection of scientific, or perhaps any, management practices. They complain that by wilderness designations they are prevented from managing the forest. Yet this confuses means and ends. The dispute is not over the fact of management *per se*, but over the goals of management. Are we to stress commodity goals or environmental quality goals in forest policy? Science and technology are just as essential to the latter as to the former. Their shape and form are different, and they emphasize systematic data collection and application more than machines, "software" more than "hardware." But even with "wilderness management," an increasingly used term, it is necessary to understand in detail the patterns of resource use and their impact on the ecological system of the forest. Management in this case is geared more to "people management" than "commodity management," with an eye to identifying the carrying capacity of the forest and restricting human use to those limits.

Gifford Pinchot was a pioneer in the 20th-century American

conservation movement. The current environmental movement is shaped not by tendencies which he helped to set in motion, but more by the new values of the post-World War II years and involves impulses which are contradictory to the objectives which Pinchot emphasized. Despite all this he remains an increasingly legendary figure in the history of American natural resource policy and especially is associated with the emphasis on harnessing science and technology to large-scale forest commodity management in the United States.

15

FREDERICK WINSLOW TAYLOR AND SCIENTIFIC MANAGEMENT

Gail Cooper

Even as a child Frederick Winslow Taylor was obsessed with order, making up elaborate rules for the informal games his playmates organized. That search for order characterized his own working life, but it also touched the concerns of the nation in the decades spanning the turn of the century. Immigration, urbanization, and industrial growth were revolutionizing the traditional American society of small towns and personal relationships. Along with these changes came social strife. Taylor, a mechanical engineer, attempted to find technical solutions to labor conflicts in American industry by increasing the surplus of wealth through greater efficiency. His study of order in the workplace prompted a popular craze throughout America for both personal and national efficiency. Theodore Roosevelt once stated, "The individual is first made efficient. . . . From these efficient units is built up an efficiency organization. And when we get efficiency in all our industries and commercial ventures, national efficiency will be a fact."

Taylor's system for greater industrial efficiency, called scientific management, was aimed at increasing national productivity and eliminating the conflict between workers and employers. As he explained to a congressional committee in 1911, "The great revolution that takes place in the mental attitude of the two parties under scientific management is that both sides take their eyes off the division of the surplus as the all-important matter, and together turn their attention toward increasing the size of the surplus until this surplus becomes so large . . . that there is ample room for a large increase in wages for the workman and an equally large increase in profits for the manufacturer." While he expressed sympathy for those who were overworked in grimy sweatshops, Taylor believed workers who underworked posed a much greater problem. He accused them of "soldiering," deliberately restricting their speed and output. He argued that if workers increased their production the resulting surplus would be sufficient to give workers a greater daily wage and management larger profits. This fair day's wage for a fair day's work would benefit both workers and employers. Taking a fairly simple view of human motivations, Taylor believed scientific management gave workers what they chiefly wanted, high wages, and employers what they most desired, a low labor cost. Rather than simply exhorting employees to work harder, management should employ "scientific" methods to establish a standard for a proper day's work.

Taylor was right that some workers deliberately limited their production. Workers called this a "stint," the amount of work that they had informally determined to be a fair day's work. The stint was an average of the workmen's abilities and protected them from unreasonable speedups by the boss. To exceed the stint for the sake of better pay or to please the boss was considered selfish and simply a way of advancing an individual's self-interest at the expense of his buddies. Workers characterized those who did so as "hoggish" or "unmanly." Whether the stint represented a reasonable amount of work or not was irrelevant to Taylor. Fundamental to his system was the claim that management, not workers, should control the workplace. This conflicted with traditional work patterns in many industries where skilled craftsmen supplied the technical know-how to run the workshop and management concentrated on the commercial side of the business. In these industries, craftsmen enjoyed a large degree of autonomy over the techniques of production and even hired and paid their own assistants. That monopoly over technical knowledge gave skilled workers considerable power over their own working conditions. In contrast, Taylor argued that management should control the methods of production and that there was no limit to management's right to know.

Taylor drew upon his own experience as both a workman and a manager to give credibilty to his system. Born into a wealthy Philadelphia family in 1856, Taylor was expected to become a lawyer like his father and in 1874 passed the Harvard entrance

Frederick Winslow Taylor

Source: Samuel C. Williams Library, Stevens Institute of Technology, Hoboken, New Jersey

165

examination with honors. He must have felt some ambivalence about his family expectations, however, for the same year he began to complain of failing eyesight, abruptly withdrew from school, and took a position as a journeyman machinist. Although his eyesight was soon "restored," Taylor continued training as a machinist, earning nothing his first year, $1.50 a week in the second and third years, and $3.00 a week in his fourth. However modest his position at work, Taylor never totally rejected the privileges of the class he was born into. During the four years of his apprenticeship, he played cricket at the Young America Cricket Club, sang with a choral group, and participated in amateur theatricals like many young men of the Philadelphia upper class. When he joined the Midvale Steel Works in 1878 as a common laborer, one owner, E. W. Clark, was the father of his tennis partner; the second, William Sellers, was a family friend whom Taylor always addressed as "Uncle William." While at Midvale, Taylor took a home course in engineering and graduated from Stevens Institute of Technology.

It was at Midvale that Taylor was first appointed to a supervisory position as gang boss of the machinists. As a machinist Taylor had worked the stint like the others; as a supervisor he was determined to increase production despite worker resistance. "Those men were my personal friends," Taylor said later, "but when we went through the gate of that place we were enemies." The antagonisms between labor and management were intense throughout American industry, fueled in part by the obvious contrasts between the personal fortunes amassed by a few industrialists like Andrew Carnegie and the grim conditions of the urban working poor. Middle-class reformers who were part of the Progressive movement brought public attention to poor working conditions in muckraking exposés and in fictional works like Upton Sinclair's *The Jungle*. But for some people sympathy turned to fear as workers unionized and strikes occasionally led to violence. While workers could often count on the sympathy of local law-enforcement officials in their communities, large companies could call upon state and federal government troops to break strike efforts. The effect was of a nation at war with itself; national concern about "the labor problem" led to the creation of the Department of Labor in 1913.

By the turn of the century, an increasing number of engineers were finding themselves in management positions mediating between the interests of workers and industrialists. Both the growing scale and the technical complexity of industrial concerns pushed the formation of a professional managerial class. The railroads pioneered modern management techniques when they adapted the army's system of line-and-staff organization to coordinate their far-flung enterprises. In 1886 Henry R. Towne suggested to the American Society of Mechanical Engineers (ASME) that engineers could make important contributions to better management of industry. His suggestion was timely; in the forty years between 1880 and 1920 the engineering profession increased by almost 2,000 percent, from 7,000 to 136,000 members. It is not surprising that Taylor first presented a part of his system of scientific management to the Society in an 1895 paper titled "A Piece Rate System Being a Step Toward a Partial Solution of the Labor Problem."

Taylor's system included a reorganization of the factory, a timed study of individual tasks, and a piece rate rather than hourly wage. Work was routed through a central planning office responsible for supplying raw materials, the appropriate tools, and for directing each individual task. It was no longer part of a machinist's job, for example, to locate his own materials or to clean and sharpen his own tools. Once the factory was rationalized, efficiency experts used stop watches to time workmen at each job, stopping the clock to exclude any unnecessary or repetitive actions, and establishing the most efficient time and method possible for each task. Workers were given a written instruction card on how to do each task. A differential piece rate rewarded those who exceeded the standard by paying more for each piece produced than was paid if the old level of production was maintained, and under Taylor's original plan, those workers who failed to produce up to the standard were paid at such a low piece rate that they were encouraged to quit and find other work.

By standardizing working conditions and by timing each task in the factory, Taylor believed he had turned management into a science. He argued that there was "one best way" to do every kind of work and that it was best determined by management. Workers were incapable of directing their own work "either through lack of education or through insufficient mental capacity." Critics charged that by separating the planning of each task from its execution, Taylorism turned workers into virtual human machines.

That attitude toward workers can be seen in Taylor's attempts at improving efficiency at the Bethlehem Steel Company. Taylor left the Midvale Steel Company in 1890 to become a management consultant. After several unsuccesful jobs in which he argued with top company officials over the extent of the engineering expert's power in the factory, in 1898 a former supervisor at Midvale paved the way for Taylor to come to Bethlehem Steel Company to institute a piece-rate system. Taylor took with him several of his experienced associates including Carl Barth; although stormy, the years at Bethlehem were fruitful for Taylor. He combined a reorganization of the steel works with experimental tests of his idea about labor. Traditionally machinists thought the best method for running the lathe was one which preserved expensive cutting tools for the longest useful life. Taylor's goal was to find a method that removed the most metal in the shortest time. He conducted systematic tests on machine lathes and discovered that standard steel-cutting tools that contained small amounts of tungsten and chromium could remove 200 to 300 percent more metal when run at high speeds that heated the cutting steel to near the melting point. With Barth's mathematical skill Taylor assembled slide rules that told machinists exactly how to set their machines for each cut. This discovery of "high-speed steel" earned Taylor considerable attention in engineering circles and in 1906 he was elected president of the ASME. His experience with high-speed steel also supported Taylor's conviction that management, not the workforce, was best able to control even skilled work in the factory.

Other tests on simple tasks reinforced his belief. At Midvale Steel Works Taylor's compulsive systematization led to years of unsuccessful experiments on ways to measure work effort. "The greatest obstacle to harmonious cooperation between the workmen and the management lay in the ignorance of the management as to what really constitutes a proper day's work for a workman," Taylor wrote. Carl Barth's mathematical analysis of the data convinced Taylor that maximum work could be achieved by regulating the percentage of time a worker spent at work and at rest. In one of the simplest jobs at Bethlehem workers loaded pig iron, rough cylinders of iron, onto a railroad car for ten hours a day at the rate of 12.5 long tons of iron per man per day. Each man picked up a

92 pound pig, walked up an inclined ramp, and dropped the chunk of metal onto the bed of the car. From his theories on rest and work percentages Taylor calculated that a first-rate pig-iron handler was capable of loading 47.5 long tons of pig iron (106,400 pounds) a day, but that seven out of eight men would not be physically capable of such work. Selecting the proper workmen for each task was important to Taylorism. Equally important was bargaining with individuals rather than groups of workers, for Taylor wanted to substitute individualism for the group ethic represented by the stint.

For his experiment he singled out a man he called Schmidt, who had ambition and stamina and who often trotted home to work on a house he was building for himself on a small piece of ground that he had bought with wages of $1.15 a day. Taylor knew no worker would agree with a supervisor's suggestion that with more effort he could increase his work from 12.5 tons to 47.5 tons per day, so he illustrated the kind of approach that was necessary with workers. He took Schmidt aside and after some preliminary questions told him, "What I want to find out is whether you are a high-priced man or one of these cheap fellows here. What I want to find out is whether you want to earn $1.85 a day or whether you are satisfied with $1.15, just the same as all those cheap fellows are getting." When Schmidt agreed he wanted to earn $1.85 a day, Taylor indicated a pile of iron and told him that to earn his higher wage he would have to load all of it the next day under the direction of an efficiency expert. Taylor informed him roughly, "Well, if you are a high-priced man, you will do exactly as this man" (pointing to the time-study man) "tells you to-morrow, from morning to night. . . . Now a high-priced man does just what he's told to do, and no back talk. Do you understand that? When this man tells you to walk, you walk; when he tells you to sit down, you sit down, and you don't talk back at him. Now you come on to work here to-morrow morning and I'll know before night whether you are really a high-priced man or not."

Under that kind of supervision, Schmidt loaded 47.5 tons of pig iron, earning 60% higher wages than ever before for 380% more work. Taylor saw no inequity in this development, focusing instead on the fact that both the workman and the company were better off. Nor did he have any regrets about Schmidt's loss of control over even the minutia of his job. Instead Taylor felt confirmed that he had discovered a new science. In his study he concluded, "Therefore the workman who is best suited to handling pig iron is unable to understand the real science of doing this class of work. He is so stupid that the word 'percentage' has no meaning to him, and he must consequently be trained by a man more intelligent than himself into the habit of working in accordance with the laws of this science before he can be successful."

Taylor's hopes to establish the engineer as the practitioner of the new science of management with benign dictatorial powers in the factory inevitably collided with the power of traditional management. At the Bethlehem Steel Company, Taylor's insistence on total control over production came at the expense of both workers and top management and in 1901 he was fired. Yet the public found his ideas of solving the "labor problem" by "scientific impartiality" to be powerfully attractive. After leaving Bethlehem Steel, Taylor retired from engineering and devoted himself to promotion of his system of management. Interest in his system spread beyond industry to the general public in 1910 when attorney Louis Brandeis argued at a rate-increase hearing that inefficient management by the railroads ultimately cost the consumers and that the adoption of scientific management could save the railroads $1 million a day. Such a sensational claim captured the public's attention. In 1911 Taylor published *The Principles of Scientific Management* to explain scientific management, and in 1912 the Taylor Society was formed by his admirers to perpetuate his work. Many of Taylor's followers and associates, including Carl Barth, Morris L. Cooke, Henry Gantt, and Frank and Lillian Gilbreth, became prominent figures in the efficiency mania that swept the nation.

Taylor's concern with efficiency was given a slightly different twist by Frank and Lillian Gilbreth, who worked as a team to introduce their own brand of scientific management to industry. Frank (1868–1924) had begun his working life as a bricklayer and his interest in speed and efficiency led him to a study of motion. He began with the bricklayer's craft, inventing a special scaffolding at waist height that eliminated the need for bricklayers to bend over to pick up each brick and at the same time teaching them to work with both hands. He became a succesful contractor in Boston and in 1904 published *Field System*, a description of his methods. After 1912 he gave up contracting and, together with his wife,

Lillian Gilbreth poses for a study of movement.

Source: National Museum of Natural History, Smithsonian Institution

devoted his energies to motion study and scientific management.

The Gilbreths's contribution was to break down work into basic motions and to eliminate unnecessary movements. These basic motions Frank named "therbligs," which he found unnecessary to point out was Gilbreth spelled backward. His analysis of movement became so fine that he used motion-picture cameras with a finely graduated clock appearing in the lower corner to record the movement of workers. Unlike Taylor, the Gilbreths's goal was not only to save time but to eliminate worker fatigue, to "make it easy for a man to work hard." Frank Gilbreth found Taylor's scientific management to be congenial to his own interests and his personal admiration of Taylor was spoiled only when Taylor refused to recognize micromotion studies as different from his own time study.

Frank Gilbreth's death in 1924 left his wife Lillian Moller Gilbreth (1878–1972) as the sole support of their eleven surviving children. As a partner in the Gilbreth consulting firm, Lillian's strength was the use of psychology in management. An Oakland, California, native who was the first female graduation speaker at

171

Frank and Lillian Gilbreth set up a time-and-motion study of a clerical worker.

Source: National Museum of Natural History, Smithsonian Institution

the University of California at Berkeley, she earned a doctorate in psychology from Brown University in 1915. As a woman in engineering she encountered some barriers to full participation in her profession. She became an early member of the newly founded American Management Association, but despite her active involvement in industrial engineering, the more traditional American Society of Mechanical Engineers twice barred her from entering the all-male Engineering Club in New York to attend meetings. Her always-busy schedule combined with the Gilbreth enthusiasm for efficiency, meant that their large household became a laboratory for domestic efficiency measures that were humorously portrayed in the book *Cheaper By the Dozen.*

After Frank's death Lillian continued to consult with corporate clients; however, she turned more and more to teaching management techniques and joined the faculty at Purdue University in 1935. She devoted considerable time to domestic efficiency. Her emphasis on an improved physical layout and the elimination of fatigue in the home made substantial contributions to the field of home economics. Most attempts to apply Taylor's scientific management to the household, however, foundered on the substantial differences between the home and the factory as a workplace. Separating the planning and the execution of each task was a questionable exercise when the housewife represented both labor and management, selecting the proper workman for each task futile when the housewife was solely responsible for a wide variety of chores, and motivating workers with piece rates pointless as long as housework remained an unpaid occupation. Recommendations to apply scientific management to the home urged housewives to save time by more efficient planning and to use that time for individual pursuits, just as workmen were encouraged to excel at their jobs for individual gain. But most women appear to have reinvested any spare time gained into household tasks that would raise the standard of living for their families.

While Taylor's primary interest in efficiency was to elminate soldiering, those who followed him answered the question, "Efficiency by what measure?" in a different way. Henry L. Gantt (1861–1919) was an engineering graduate of Stevens Institute of Technology who worked with Taylor at both Midvale and Bethlehem. He was a Southerner by birth and tradition whose lifelong hero was "Stonewall" Jackson. The Civil War had created financial disaster for his family and he was educated at a military-style boarding school for poor boys just outside of Baltimore. He first met Taylor at the Midvale Steel Works. In 1987 Taylor attempted to reorganize the Simmonds Rolling Machine Company but met strong resistance by middle management. When the general manager, all foremen and assistant foremen, the salesmen, and the head men in the office resigned on three days' notice, Taylor installed Gantt as superintendent. From Simmonds both men went to the Bethlehem Steel Works.

Gantt was temperamentally kinder and more flexible than Taylor, and at Bethlehem he modified Taylor's differential piece rate with the task-and-bonus system. Under this system, workers who failed to achieve their task for the day were simply paid by the hour rather than at Taylor's very low piece rate, while "efficient" workers were paid a bonus. Gantt's later willingness to compromise with his engineering clients made Taylor wonder if Gantt really understood scientific management and, like his relationships with others, Taylor's friendship with his protégé soured after 1908 when Gantt developed independent ideas about workers' training. As his thinking about industrial management evolved, Gantt was as critical of inefficiency in top management as Taylor was of labor. His most important technical contribution, a graph showing machine production known as the Gantt chart, became part of his attempt to establish standards for management analogous to time studies for workers.

When workers failed to achieve their bonuses Gantt was always interested in finding out why; his investigations often led him to discover a failure in management. His critique of management shifted from a concern with increased profits to a deepening conviction that factories should be run for the public good rather than simply for private profit. Just as Taylor aimed to benefit workers and employers by maximizing production and profits, Gantt extended that concern to include economical production for the benefit of the consumer. He deplored the tendency of inefficient companies to allow the machinery to stand idle and to pass the cost of this expense on to the customer. Industry worked at only a small percentage of its capacity because shortages resulted in higher prices and higher profits for factory owners. To discourage the industrial capacity of the country being run for the private profit of a few, Gantt advocated a new system of bookkeeping in which the cost of idle machinery was not added to the product but instead taken directly out of profits.

In contrast to his disillusionment with the selfish concerns of financiers and businessmen, he envisioned an important role for the engineer not only in industry but also in society. Influenced by economist Thorsten Veblen, at a meeting of the ASME in 1916 he and thirty-three others formed a group called the New Machine, dedicated to increasing the purchasing power of a day's pay in New York City by readjusting the social and political approaches to the challenges of industrial society. However, Gantt did not have the personality for politics. As his biographer observed, "His inherent honesty, lack of tact, fearless straightforwardness, brusque expression, and loneliness, because of lack of friends, are

174

not the human qualities which make for the success of the politician." The New Machine was pushed aside by the concerns of World War I.

While many of Taylor's followers expanded both the techniques and the scope of his system, Carl Barth remained the most orthodox of the scientific-management experts and was one of Taylor's favorite lieutenants. In 1909 Taylor recommended that Barth bring scientific management to the U.S. Federal Arsenal in Watertown, Massachusetts. The arsenal became a battleground between exponents of scientific management and organized labor. Taylor was anxious to gain implicit approval of scientific management from the federal government, while the union's resistance focused on preventing the introduction of the stop watch. When Barth's assistant conducted his first time study in the foundry, workers surreptitiously timed the man under study as well; the resulting discrepancy fueled a strike among the foundrymen the next day. The expert was timing an ideal process, not the actual work time, but had to admit that he knew little about the technical requirements of the task and that he had settled on a time based on what he felt was appropriate.

The "science" of Taylor's system seemed ultimately to have been based on subjective judgment cloaked in precise measurement, and even supporters who believed the system to be effective resisted its claims to scientific precision. When Taylor asked the American Society of Mechanical Engineers to form a special committee to evaluate the scientific basis of Taylorism in 1912, they returned a report against Taylor citing the imprecision of terms like "first-class workman" and the substantial allowances that had to be added to each task time to account for unavoidable delays. In reality Taylor was trying to measure the unmeasurable, and the intolerance of his "one best way" was a measure of the man as well as the system.

After congressional hearings on the new system, workers succeeded in blocking Taylorism at Watertown Arsenal by lobbying their congressmen to outlaw the use of Taylorism on some government contracts. The passage of congressional legislation against the use of scientific management left Taylor discouraged. He lost heart for his work and his health declined. He suffered from a lingering cold; on a trip to visit his sister in Philadelphia, his cold turned to pneumonia and he died on March 21, 1915, at the age of fity-nine.

Taylorism was only a partial success during its founder's lifeime due to Taylor's difficult personality and to the inherent flaws in the system. Scientific management was seldom adopted in its entirety, and the comprehensive installation at the Tabor Manufacturing Company, which Taylor used repeatedly as a living example of his methods, was notable for its rarity as well as its effectiveness. In addition, claims of scientific precision crumbled during Taylor's lifetime and with them the force of its moral imperative. Taylor had claimed that the discipline of the task-and-bonus system improved the moral character of the worker and led to sobriety, ambition, and thrift. That view of the relationship between production and workers' welfare was inverted by a more persuasive group of reformers who provided workers with subsidized housing, savings plans, and leisure activities in the belief that material well-being was the precursor to increased production.

However, the highly publicized victory of labor over Taylorism did not stop the spread of scientific management. Taylorism was remarkably influential both at home and abroad where it seemed to many to be an integral part of America's industrial success. In addition, scientific management experts like Lillian Gilbreth launched the new field of industrial psychology and personnel management. Scientific management was also long lived. The increasing technical complexity of large-scale industry from 1880 to 1920 that encouraged the growth of systems of management remained a permanent feature of industrial America.

16

HENRY FORD AND THE TRIUMPH OF THE AUTOMOBILE

James J. Flink

By 1908, the year in which Henry Ford introduced his Model T and William C. Durant founded General Motors, the automobile, in the words of technology historian John B. Rae, "was already potentially what it would become in fact—an incredible item of mass consumption." By 1927, when the Model T was withdrawn from production, the automobile industry was America's leading industry in value of product; there was an automobile registered for every 4.5 persons in the United States; and 55 percent of American families owned cars. A presidential commission surveying contemporary changes in American life concluded about the automobile in 1933: "It is probable that no invention of such far reaching importance was ever diffused with such rapidity or so quickly exerted influences that ramified through the national culture, transforming even habits of thought and language."

More recently, Father R. L. Bruckberger, a French observer of American life, has concluded in his *Image of America* that Henry Ford's innovation of the Model T, the moving belt assembly line, and the five-dollar, eight-hour day have had more important consequences than Lenin's socialist revolution. Currently the automobile industry directly provides one out of every six jobs in the

United States; there is an automobile registered for every 2.25 persons; and 83 percent of American families—all except the very poor—own cars. The automobile makes possible and has come to symbolize much of the American way of life.

Ironically, "the automobile is European by birth, American by adoption." Until shortly after the turn into the 20th century, Europeans were about a decade ahead of Americans in development of the gasoline-powered car. The automobile was pioneered to the stage of commercial feasibility as early as 1887 by Karl Benz, a German manufacturer of stationary gas engines, and the mechanical arrangement of the modern motorcar was innovated in 1891 by the French engineer-entrepreneur Emile Constant Levassor. In 1895 Levassor drove one of his cars over the 1,170-kilometer course of the Paris-Bordeaux-Paris race at the then incredible speed of just over 24 kilometers per hour, with the longest stop for servicing being only 22 minutes. This demonstrated that the replacement of the horse by a motor was more than an idle dream and touched off a flurry of automotive activity in the United States, where Levassor's achievement was given extensive coverage in the newspapers. Although a handful of American inventors had managed to build experimental cars by the mid-1890's, their machines were primitive by European standards and were not being offered in quantity for sale. In contrast, by 1895 several French and German automobile manufacturers were issuing catalogues, and automobiles were already a common sight in the streets of Paris.

Commercial manufacture of motor vehicles in the United States was initiated in 1896 by Charles E. and J. Frank Duryea, two Springfield, Massachusetts bicycle mechanics credited with building the first successful American gasoline-powered car in 1893. Other early experimenters, including Henry Ford, soon entered the competition, and some 30 American automobile manufacturers produced an estimated 2,500 motor vehicles in 1899, the first year for which separate figures were compiled for the automobile industry in the *United States Census of Manufactures*. Despite the fact that cars were then sold on a cash-on-delivery basis and left much to be desired in performance, demand for motorcars in the United States far exceeded the supply. The early automobile manufacturers operated in an unprecedented seller's market for an expensive item. Consequently, by 1910 automobile manufacturing had leaped from 150th to 21st in value of product among American industries and had become more important to the national economy on all measurable criteria than the wagon and carriage industry. By 1910, 458,500 motor vehicles were already registered in the United States, making America the world's foremost automobile culture.

There are many reasons why, in contrast with Europe, an automobile culture developed so rapidly in the United States. To begin with, the volume production of standardized commodities became well established early in our industrial history. Our abundance of natural resources, combined with a chronic shortage of labor, resulted in low costs for raw materials and the mechanization of industrial processes, which necessitated the standardization of products. In addition, the absence of tariff barriers between the states encouraged sales over a wide geographic area. Most important were our higher per capita income and more equal income distribution relative to those in European countries. It is significant, for example, that Morris Motors, the most important British automobile manufacturer, did not install a moving assembly line until 1934—two decades after it had been innovated at Ford. The income distribution in Great Britain fixed the demand for cars there at too low a level to justify the investment. Because of these differences between Europe and the United States, the European pattern of small-scale, individualized production of motor vehicles stood no chance of becoming characteristic of the American automobile industry.

The automobile was enthusiastically received by Americans from its introduction, and shortly after the turn of the century predictions became commonplace that a utopian ''horseless age'' would dawn with the imminent development of the low-cost, reliable car. Even before the introduction of the Model T in 1908, most Americans anticipated that mass personal automobility would soon become a reality. The bicycle had made Americans conscious of the possibilities of individualized, long-distance highway transportation, and the crest of the bicycle movement in the 1890's coincided with the climax of several decades of agrarian unrest that had singled out as a prime target the abuse of monopoly power by the railroads. Farmers and city consumers alike began to perceive cheaper highway transportation as an alternative. Vast areas of the American continent remained inaccessible by rail or highway, and most Americans still lived on isolated farms or in poorly interconnected rural villages. They wanted better access not only to markets, but to the social and cultural amenities that only the metropolis provided, particularly to better medical care and education. In cities the antiseptic automobile was viewed as an answer to the health problems arising from ever-present manure heaps and the traffic congestion caused by overworked dray horses dropping dead in their tracks. The urban middle class, imbued with the Jeffersonian "agrarian myth," sought escape to suburban residences through automobility. Municipal governments hard pressed to meet the mounting costs of building adequate mass transit rail systems saw an answer in the widespread adoption of motor vehicles. Countless early demonstrations convinced people that the automobile was more economical, more reliable, and safer than the horse-drawn vehicle. And, perhaps most important, the motorcar offered our individualistic, migrant population the promise of greatly expanded personal mobility and freedom of choice in residence, business location, and the pursuit of leisure-time activities.

Henry Ford was by far the most successful of the many early automobile manufacturers who attempted to build a low-cost, reliable car for this developing mass market. Born on a Michigan farm in 1863 of middle class parents, Henry Ford early developed a life-long aversion to the drudgery of farm labor. As a youth, he loved to tinker with machinery, and, while employed as an engineer at Detroit's Edison Illuminating Company, he built his first experimental automobile in 1896. During two unsuccessful attempts to enter automobile manufacturing, Ford gained a national reputation as a designer and driver of racing cars. Then at the age of 40, with new backers, Ford formed the Ford Motor Company in Detroit, Michigan, on June 16, 1903, to produce cars in the intermediate-price range. The Ford Motor Company started in business with paid-in capital of only $28,000, a dozen workmen, and an assembly plant just 250 by 50 feet (76 by 15 meters). So little capital was needed because Ford like almost all other early automobile manfuacturers jobbed out the manufacture of components to scores of independent suppliers. Once the design of his car was established, the early automobile manufacturer became merely an assembler of major components and a supplier of finished cars to his distributors and dealers. As a consequence of these easy conditions of entry—low capital requirements combined with the promise of high profits— some 515 separate firms had entered automobile manufacturing by the time Ford came out with his Model T in 1908. Competition was intense: firms that made technological blunders or failed to adjust to changes in demand were weeded out with Darwinian ruthlessness. The vast majority of Ford's early competitors failed. Why did he succeed so brilliantly?

Ford was not the first automobile manufacturer to attempt to produce in volume low-cost, reliable cars. That distinction belongs to Ransom E. Olds, who initiated volume production of his $650 curved-dash Oldsmobile in 1901, while Ford was still building racing cars. But both Olds and Thomas B. Jeffery, who followed Olds into volume production with his $750 to $825 Ramblers in 1902, attempted to mass-produce cars that were soon outmoded. The surrey-influenced design of the curved-dash Oldsmobile had inherent mechanical weaknesses and, consequently, was rapidly abandoned by most manufacturers. The typical gasoline automobile of 1908 bore little resemblance to the horseless carriage of 1900. Numerous improvements meant that the 1908 gasoline automobile was a fairly reliable family car. The problem remained prior to the Model T, however, of making such cars available at prices the average family could afford, by cutting manufacturing costs or reducing unit profits without sacrificing the quality of the product.

At the time the Ford Motor Company was organized, the automobile industry was dominated by the Association of Licensed Automobile Manufacturers (ALAM), a select group of makers of gasoline automobiles who were attempting to monopolize automobile manufacturing in the United States through control of a patent on the gasoline automobile that had been awarded in 1895 to George B. Selden, a New York patent attorney. The ALAM companies (which accounted for about 80 percent of the industry's total output prior to 1908) were in the main committed to maintaining high unit profits through producing high-priced cars for a limited market. The ALAM tried to set production quotas and to freeze new entrances into automobile manufacturing. Henry Ford was denied a license to manufacture gasoline automobiles under the Selden patent on the ground that he had not demonstrated his competence, and when Ford persisted in producing cars, the

ALAM immediately brought a lawsuit against him for infringement of the Selden patent. The suit was ultimately decided in Ford's favor in 1911 and the ALAM disintegrated. Meanwhile, Ford turned the suit into a clever public relations campaign by contrasting his own status as a pioneer automotive inventor and struggling small businessman with the image of the ALAM as a group of powerful and parasitical monopolists.

Like many other automobile manufacturers who operated without licenses from the ALAM, Henry Ford was increasingly committed to the volume production of light, low-priced cars. His $600 Model N was one of the better-designed and better-built cars available at any price in 1906. The *Cycle and Automobile Trade Journal* called the Model N "the very first instance of a low-cost motorcar driven by a gas engine having cylinders enough to give the shaft a turning impulse in each shaft turn which is well built and offered in large numbers."

Encouraged by the success of the Model N, Henry Ford was determined to build an even better low-priced car. At $825 for the runabout and $850 for the touring car, the four-cylinder, 20-horse-power Model T was first offered to dealers on October 1, 1908. Ford's advertising claim about the Model T was essentially

Henry Ford stands beside the first automobile he built in 1896, while his son Edsel inspects the ten millionth Ford car, a Model T, which Ford introduced in 1908.

Source: Ford Motor Company

182

correct: "No car under $2,000 offers more, and no car over $2,000 offers more except the trimmings." Committed to large-volume production of the Model T as a single, static model at an ever-decreasing unit price, the Ford Motor Company innovated mass-production techniques at its new Highland Park plant, which permitted prices to be reduced by August 1, 1916, to only $345 for the runabout and $360 for the touring car. Model T production increased form 32,053 units in 1910 to 734,811 units in 1916, giving Ford about half the market for new cars in the United States at the outbreak of World War I.

The Model T was the archetype of a uniquely American mass-produced gasoline automobile. Compared with the typical European touring car, the American-type car was significantly lower priced, was much lighter, had a higher ratio of horsepower to weight, and was powered by a larger-bore, shorter- stroke engine. In addition to the increasingly greater emphasis that American manufacturers gave to producing cars for a mass market, these characteristics of the American car resulted from the lower price of gasoline and the absence of European horsepower taxes. The sacrifice of fuel economy for greater engine flexibility made it possible to build cars that could more readily negotiate steep grades and wretched roads and were easier to drive because they required less frequent shifting of gears. The American car, designed for the average driver rather than for the professional chauffeur, was more apt to withstand abuse and was simpler to repair than the typical European car. However, with a body that sat higher above the roadbed, the American car appeared ungainly and showed less attention to the details of fit, finish, and appointment. That the chassis of a car clear the invariable hump in the center of our unpaved, rutted roads was then more important to the American consumer than style.

The 60-acre Highland Park plant that Ford opened on January 1, 1910, to meet the huge demand for the Model T possessed an unparalleled factory arrangement for the volume production of motorcars. Company policy was to scrap machines as fast as they could be replaced by improved types, and by 1912 the tool department was constantly devising specialized new machine tools that would increase output. By 1914 about 15,000 machines had been installed. Time and motion studies led in 1912 to the installation of continuous conveyor belts to bring materials to the assembly lines.

By the summer of 1913 magnetos, motors, and transmissions were assembled on moving lines. After the production from these sub-assembly lines threatened to flood the final assembly line, a moving chassis assembly line was installed, reducing the time of chassis assembly from 12 and a half hours in October to two hours and 40 minutes by December 30, 1913. The moving lines were at first pulled by rope and windlass, but on January 14, 1914, an endless chain was installed. "Every piece of work in the shop

One of the first applications of the moving assembly line was this magneto assembly operation at Ford's Highland Park plant in 1913. Magnetos were pushed from one workman to the next, reducing production time by about one-half.

Source: Ford Motor Company

moves," boasted Henry Ford in 1922. "It may move on hooks or overhead chains going to assembly in the exact order in which the parts are required; it may travel on a moving platform, or it may go by gravity, but the point is that there is no lifting or trucking of anything other than materials."

These mass production techniques greatly decreased unit production costs, which Ford passed on to the consumer in the form of lower Model T prices, to a low of only $290 for the coupe by 1927. Soon applied to the manufacture of many other items, these mass production techniques significantly increased the standard of living of the average American family and shifted the American economy from a production-oriented economy of scarcity to a

184

The body drop was the final step in assembling the Model T at Highland Park. Eventually Ford sped up the assembly process so that a Model T was completed every ten seconds.

Source: Ford Motor Company

consumption-oriented economy of affluence. Recognizing ahead of his contemporaries that mass production necessitated mass consumption and that workers had to have the means to purchase what the machine produced, in 1914 Ford inaugurated for his workers the five-dollar, eight-hour day, which more than doubled wages for a shorter work day. Mass production also created a new class of semi-skilled workers, who needed neither substantial training nor physical strength. New remunerative employment opportunities were thus opened to the immigrant, the black migrant to the northern city, women, the physically handicapped, and the educable mentally retarded. Unable to compete with Ford's progressive lowering of Model T prices through mass production, the makers of moderately priced cars in 1916 began to sell cars on credit, and by 1926 installment sales accounted for about two-thirds of all new car sales. Led by the automobile industry, the credit buying of expensive items thus became an established part of American life by the mid-1920's. Finally, the huge outlays of capital necessitated by mass production for gigantic specialized plants and complex machinery were beyond the means of the small automobile producer. Inevitably, automobile manufacturing changed from an industry characterized by intense competition among many small firms into an oligopoly by the late 1920's, with Ford, General Motors, and Chrysler accounting for about 80 percent of the industry's total output.

By the mid-1920's automobile manufacturing ranked first in value of product and third in value of exports among American industries. The automobile industry was the lifeblood of the petroleum industry, one of the chief customers of the steel industry, and the biggest consumer of many other industrial products, including plate glass, rubber, and lacquers. The technologies of these ancillary industries, particularly steel and petroleum, were revolutionized by the new demands of motorcar manufacturing. The construction of streets and highways was the second largest item of governmental expenditure during the 1920's. The motorcar was responsible for a suburban real estate and construction boom (which has continued into the present, permitting most American families to live in single-family residences) and for the rise of many new small businesses, such as service stations and tourist accommodations. With the new mobility of the population, business locations and residential patterns became more decentralized, and the Sunday drive to the country and the annual automobile vacation to distant national parks became commonplace. The Model T permitted farmers to shop in the city, killing off the crossroads general store and the mail order houses. Conditions of rural life and farm labor were also radically altered by the school bus, which replaced the one-room school with the consolidated, graded school, and by the small farm tractor—exemplified by the Fordson—which diminished the need for farm workers while greatly increasing agricultural productivity and alleviating much of the back-breaking drudgery of farm labor. (The larger economic units required for efficient utilization of the tractor and the other mechanized equipment its use encouraged, however, ultimately made the small family farm obsolete.) Long-distance trucking, developed during World War I, opened up the Pacific Coast and the Southwest to commercial exploitation and knit regional economies more tightly together into a national economy. Noting the central role of automobility to changes in the American society and economy in the 1920's, the eminent historian Thomas C. Cochran has concluded: "No one has or perhaps can reliably estimate the vast size of capital invested in reshaping society to fit the automobile . . . . This total capital investment was probably the major factor in the boom of the 1920's and hence in the glorification of American business.''

By the end of the 1920's the automobile had triumphed in the United States. The American society and economy had been restructured by the motor vehicle, and automobile ownership had been extended as far as the income distribution of the day would permit. The year that the Model T was withdrawn from production, 1927, was also the first year that replacement demand for new cars exceeded demand from first-time owners and multiple car purchasers combined. In 1929, the last year of the automobility-induced boom, the 26.7 million motor vehicles registered in the United States travelled an estimated 198 billion miles. During the Great Depression of the 1930's although automobile sales slipped drastically, the miles travelled annually by motor vehicle in the United States continued to increase, and automobility proved to be one of the most depression-proof aspects of American life. However, World War II brought a halt to passenger car production and great curtailment of automobile use through the rationing of gasoline and tires. Consequently, the automobile industry's record 1929 output of 5.3 million units was not again equalled until 1949.

With the triumph of the automobile in America came the end of the Model T era. As the need for basic transportation was met by the Model T and rural roads were vastly improved, consumers became more comfort and style-conscious. Owners whose first car was a utilitarian Model T tended to trade up to larger, more powerful, and more stylish cars. Henry Ford failed to keep abreast of this change in demand, and his competitors at General Motors and at Chrysler not only surpassed him in product design but by the mid-1920's caught up to Ford in the efficiency of their assembly lines. Except for minor face liftings and the incorporation of such basic improvements as the self-starter and the closed body, the Model T remained unchanged long after it was outmoded. By 1927 second-hand cars of more expensive makes in excellent condition could be bought for about the same price as an obsolete Model T, and an annually restyled and far better equipped new Chevrolet cost only $200 more. In 1927 Chevrolet first surpassed Ford in sales. And by 1936 the Ford Motor Company had declined to third place in sales of passenger cars, with 22.44 percent of the market versus 43.12 percent for General Motors and 25.03 percent for Chrysler. The Ford Motor Company continued to decline until its aging founder turned over its presidency to his grandson, Henry Ford II, in 1945.

During the 1920's the achievements of American capitalism were most prominently symbolized by Henry Ford, who was idolized not only by the common people of the United States but throughout the world. Ford received several thousand letters a day from ordinary people who were impressed with his accomplishments, and more was written about Ford during his lifetime (most of it adulatory) than any figure in American history. Even peasants in remote Asian villages knew about his Model T. And in the Soviet Union the world's leading industrial capitalist was hailed as an economic innovator second only to Lenin for his contributions to mass production and for the Fordson tractor.

Actually, as Reynold M. Wik, John Kenneth Galbraith, and others have shown, the success of the Model T was the result of a group effort by an outstanding entrepreneurial team at the Ford Motor Company, not the product of Henry Ford's individual genius. And, contrary to popular myth, Henry Ford did not invent mass production: its constituent elements were all parts of an evolving American manufacturing tradition, and many Ford employees contributed the key ideas that led to its implementation and perfection at the Ford Highland Park plant. After 1914, however, Ford Motor Company public relations policy was to promote Henry Ford, not the company, and to downplay the contributions of others to the success of the mass-produced Model T. The Ford mystique was thus in large part based on myths, which Henry Ford deluded himself into believing and unfortunately encouraged.

By the time of his death on April 7, 1947 Henry Ford's reputation had been tarnished by the blatantly anti-Semitic articles published in his *Dearborn Independent* (for which he publicly apologized in 1927) as well as by the deterioration of working conditions in his plants and his opposition to the unionization of his workers in the 1930's. Ford had bought out his minority stockholders in 1919 and after that tended increasingly toward arbitrary, one-man rule of his industrial empire. An increasing number of ex-Ford executives called this "Prussianization," and the *New York Times* in 1928 went so far as to call Ford "an industrial fascist." Ford refused to cooperate with President Roosevelt's plans for recovery from the Depression. Strokes suffered in 1938 and in 1940 left Ford's eroding mental faculties greatly impaired.

Nevertheless, Henry Ford remains a folk hero to millions of people throughout the world, who, ignoring his failings, continue to identify him as the progenitor of mass personal automobility and mass production. Henry Ford will probably remain an important figure in world history long after the names of most American Presidents are forgotten.

17

PETER L. JENSEN
AND THE AMPLIFICATION OF SOUND

W. David Lewis

During the relatively short span of less than two hundred years, human communication has been transformed by a series of technological innovations including high-speed presses and typesetting machines, the telegraph, the telephone, the phonograph, radio, television, the digital computer, and orbiting earth satellites. Of key importance in the unfolding of this process has been the growth of knowledge about electricity, which had begun to yield a flood of practical applications by the last quarter of the nineteenth century. As scientists and engineers explored the behavior of alternating current and electromagnetic waves and learned about the presence of electrons in the basic structure of the atom, the pace of change accelerated in the early twentieth century; the exciting new technology of radio served as a major focus of activity. Among the many inventions that resulted was the creation of a new and unheralded communications device, the loudspeaker.

Electronically amplified sound is such an omnipresent feature of modern society, sometimes reaching levels of volume so intense that it constitutes an environmental hazard, that it is difficult to envision a world in which it did not exist. Yet it was only in 1915 that Peter L. Jensen, a Danish immigrant, and Edwin S. Pridham, his American associate, invented the modern loudspeaker in a small bungalow on the outskirts of Napa, California. As has been true of many other inventions, its birth was accidental. Jensen and Pridham had been working on an improved telephone receiver but could not reduce its cost and size to the point at which it could be commercially exploited. Acting upon a suggestion by a relative of Jensen's wife, the two inventors transformed their brainchild into a device that, for the first time in history, electrically amplified sound far beyond any previous limits.

The invention of the loudspeaker is therefore a classic example of serendipity, the art of finding things unintentionally or by indirection. More than this, however, it illustrates the complex process of feedback that took place as accumulating scientific and engineering knowledge produced rapid changes in radio technology in the early twentieth century. The loudspeaker's technology represented a shift away from what historian Hugh G. J. Aitken has called "syntony and spark," involving methods of transmission and reception that characterized early radio, to a "continuous-wave" technology that opened up dramatic new possibilities after the turn of the century and laid the foundations of radio as we know it today.

Jensen's role in the development of the loudspeaker resulted from his employment, as a young man in his native Denmark, under one of the pioneers of continuous-wave technology, Valdemar Poulsen. The opportunity to come to America to help install continuous-wave equipment for entrepreneurs who had acquired the right to exploit Poulsen's patent rights in the United States launched a remarkable career that illustrates the process of "technology transfer" from one part of the world to another within scientific, technological, and business communities whose knowledge and activities transcended national boundaries. In time the young immigrant would help lay the foundations for modern audio technology and win both financial success and international recognition. The reputation gained by such products as Magnavox radio and phonograph sets and Jensen speakers testified to the effectiveness and importance of his work.

Jensen's career, however, was not dedicated solely to technological innovation. A sensitive and reflective person, he was obliged by the circumstances of his life to search for a new sense of personal identity a a stranger in a land that, however much it

Peter L. Jensen, circa 1930

differed from the country of his birth, he came to love. Absorbing himself in its history and traditions through the influence of a fellow engineer who became his best friend, he sought diligently to make himself over into a new self-image only to experience, in a moment of profound introspection, that he belonged to two countries and would always bear the stamp of both. His life belies a common image of engineers as persons who lack interest in emotional or cultural growth. Clearly, he was a complex man of much substance and inner depth.

Nothing in the parentage or background of Peter Laurits Jensen suggested that he would become an inventor and make significant contributions to the development of electronic technology. Born May 16, 1886, near Stubbekobing on the island of Falster in the eastern archipelago of Denmark, he seemed destined to follow the career of his father, Lods Ole Jensen, a navigator who guided vessels through the waters between Falster and the neighboring island of Lolland. As Peter later stated in his autobiography, acquaintance with the Baltic Sea, "sometimes calm and innocent looking, but often deadly in its fury," began early. At age seven he began assisting his father at the rudder of a small pilot boat that plied the turbulent Strait of Groensund while a larger sailing ship came on behind, "splashing and foaming at the bow," and sometimes threatening to collide with its tiny escort in the treacherous passage. At other times he fished for herring or helped his mother, Hansine Petersen Jensen, pick potatoes, weed beets, or tend livestock on the small plot of agricultural land that surrounded the family home. The parenting he received was austere; although he was certain of his mother's love, he could later remember no occasion on which she had ever shown him any physical affection.

Escape from this life of danger and drudgery was made possible by the precocity that Jensen displayed in his elementary schooling at the nearby village of Moseby, where he ranked at the top of his class. When he was thirteen, the senior teacher paid an unexpected visit to his parents. "No child had ever gone from his to a higher school to graduate from the state university," Jensen later wrote, "and he came to suggest that I do just that." After much debate his parents assented, and Jensen left home for a boarding school at Norre Alslev, passing the entrance examination for the University of Copenhagen three years later. He never matriculated, however, for his father died from the effects of "one of the worst storms ever to sweep down on the Baltic," forcing Peter to return home to look after his mother. Taking a job hauling fagots from a woodlot to a sawmill, his knowledge of German and English attracted the attention of a superintendent who suggested that he ought to use his education to better advantage and urged his mother to send him to Copenhagen in search of opportunity. She did so, giving him a letter to a well-known resident of that city, Lemvig Fog, whose family had at one time boarded at the Jensen home during a summer vacation.

Fog was an engineer-entrepreneur who had returned to Denmark after making a fortune in Brazil. As a financial backer of Valdemar Poulsen, he suggested that Jensen embark upon a technical career and helped him secure employment as a "mechanist apprentice" in Poulsen's laboratory. Here Jensen received training comparable to that obtained by American counterparts in what historian Monte Calvert has called the "shop culture" approach to engineering education in the United States, at that time still a viable route to a professional career despite an increasing trend toward study in an institute of technology or a university engineering department. Knowing that both Fog and Poulsen "were known as engineers and yet neither had graduated from a university," Jensen resolved to follow in their footsteps.

Poulsen was already an inventor of international consequence when Jensen entered his laboratory; at the Paris Universal Exhibition of 1900 he had won a grand prize for a device called the "Telegraphone," an early forerunner of the tape recorder. Shortly after Jansen began working for him, Poulsen won further recognition when he succeeded in amplifying electromagnetic waves generated by an arc transmitter by having the arc burn in an atmosphere of hydrocarbon vapor instead of in air. A significant advance in the development of continuous-wave technology, the "Poulsen arc" was a finicky device that permitted sharp tuning but required frequent regulation in order to work efficiently. Jensen became adept at keeping it in effective operation and was entrusted with this responsibility by Poulsen.

In 1905 Poulsen set up a radio station at Lyngby, near Copenhagen. Jensen, who had taught himself Morse code and was adept at sending and receiving radio messages, took part in its construction. Efforts to capitalize upon the transmitting capabilities of the Poulsen arc, however, were impeded by its limited range and the deficiencies of the Branly coherer, a glass tube within which metal filings between two electrodes were excited by damped waves generated by a spark transmitter, thus constituting the standard signal detector in the early days of radio. With the help of P. O. Pedersen, a professor in the Royal Technical College, Poulsen developed a new device known as the "tikker," which used a vibrating reed, two gold wires acting as an interrupter, and telephone circuitry enabling radio signals to be received through a pair of headphones or, in a high-speed version, to be registered on photographic tape for subsequent scanning. Using a crystal detector, efforts were made to transmit human speech with this equipment, a feat that was impossible with spark technology. While taking part in this project, Jensen seized upon the idea of placing a microphone in the ground circuit of the transmitter and connecting the crystal detector to a grounded tikker as a receiver, with successful results. Before serving a brief stint as a wireless operator in the Danish Navy in 1909, Jensen utilized the new system by transmitting recorded music from Lyngby to a small number of listeners, including radio operators on ships at sea.

Although Poulsen's system was less expensive and easier to switch from one frequency to another than an alternate continuous-wave system developed by Reginald Fessenden in the United States, it saw only experimental use until it attracted the attention of a group of American investors represented by Cyril F. Elwell, a native of Australia who had emigrated to California and earned a degree in electrical engineering at Stanford. Elwell had become interested in radio as a result of testing an unsuccessful system of wireless telephony developed by Francis J. McCarthy, a young inventor whose work had aroused interest among San Francisco businessmen prior to his death in a 1906 automobile accident. Seeing in Poulsen's system a potential rival to the one that Fessenden had developed in cooperation with General Electric, Elwell went to Copenhagen in 1908 and was impressed by his observations of the Danish technology, although somewhat deterred by the stiff terms demanded by Poulsen and his associates for its exploitation in America. Returning in 1909 with backing from a group of investors in Palo Alto, Elwell persuaded the Poulsen syndicate to modify its demands and take part in a California enterprise to be known as the Poulsen Wireless Telephone and Telegraph Company (PWTTC).

When Poulsen asked him to go to America to supervise installation of the Danish equipment, Jensen accepted the offer as the opportunity of a lifetime. On December 9, 1909, he left Denmark with a mechanic, Carl Albertus, who had previously lived in Chicago. Arriving in New York City after a twelve-day voyage, the two young men traveled across the continent by rail to San Francisco. Jensen marveled at the contrast between America, where he "found a most peculiar conglomeration of the best and the worst," and his native country, where everything was "clean and orderly with no extremes." The "semi-tropical brightness of California made the deepest impression upon me," he later recalled; "I had never seen palms and tropical fruit trees before, and the weather as we rolled down into the Sacramento Valley was sparkingly bright." Crossing the Bay from Oakland into San Francisco at sundown evoked a feeling that he "had arrived at the end of the world and found that end most glorious." Equipment soon arrived from Denmark and two stations were erected in the San Joaquin Valley, one in Sacramento and the other about fifty miles away in Stockton.

During completion of the Sacramento radio station, Jensen met Edwin S. Pridham, a native of Illinois who had graduated from Stanford in 1908 with a degree in electrical engineering after working his way through school by cleaning bricks in debris left over from the 1906 San Francisco earthquake. Fascinated by radio, he had taken a job with PWTTC shortly after graduating. Pridham became Jensen's "ideal of a real American." He helped Jensen polish the rudimentary English he had learned in Denmark, and taught him the lore and traditions of the United States so successfully that the study of American history became Jensen's hobby. "During the next fifteen years," Jensen later wrote of Pridham, "he never ceased in his efforts to change me from a foreigner into a true American."

Becoming inseparable friends, Jensen and Pridham soon moved to San Francisco, where a third radio station was built on the beach with two tall masts that became landmarks to local residents. Jensen later marveled that he and Pridham were oblivious to the commercial possibilities of radio broadcasting, despite the fact that amateur operators with crystal sets listened to transmissions from the new station to Sacramento and Stockton. Like PWTTC's backers, they had little sense of radio's market possibilities, seeing it mainly as a potential competitor of the telegraph and a means of ship-to-ship and ship-to-shore communication. Dominated by the Marconi Wireless Telegraph Company of America, the radio industry in the United States was characterized almost exclusively by the use of spark technology to send messages in Morse code. Although such inventors as Fessenden, John S. Stone, and Lee de Forest were developing significant new devices utilizing continuous-wave technology for transmission, tuning, and reception, they lacked effective strategies for competing successfully with the Marconi interests and the value of their work was unappreciated by the general public. De Forest in particular had a version of how "wireless" might be used to broadcast news, advertising, and music and had invented a three-element tube known as the "audion" that, along with powerful alternators developed by Fessenden, would in time make spark technology obsolete. For the present, however, only the activities of amateur operators gave any hint of the future of broadcasting; the use of radio was otherwise confined to a relatively few business institutions and governmental agencies, particularly the U.S. Navy.

As additional stations were built by Poulsen's American associates at Los Angeles and other locations, PWTTC went through a process of reorganization leading to the establishment of two new firms, the Poulsen Wireless Corporation (PWC) and its operating subsidiary, the Federal Telegraph Company (FTC). Considering his work done, Jensen decided to go back to Denmark. Before he could leave, however, he was approached by John C. Coburn, a glib promoter who had been active in selling stock for PWTTC. Having unloaded his investment in that company at a substantial profit, Coburn now represented a group of San Francisco businessmen who had failed to secure stock in FTC and were engaged in a grandiose scheme to buy the rights to exploit Poulsen's patents throughout the entire world outside the United States, hoping to acquire the remaining American rights at a later time. Believing that Jensen's influence would be useful in selling the idea to Poulsen, Coburn promised to make the young Dane chief engineer of the resulting enterprise and give him a large block of its stock.

Dubious about the project's chances for success, Jensen nevertheless agreed to go to Copenhagen on behalf of its promoters, stipulating only that Pridham must be allowed to go along. Resigning from FTC, the two friends went by way of England with the hope of getting permission to erect stations in Ireland and Canada, only to find that Poulsen had already granted an option on his English patent rights to a British syndicate. Proceeding to Copenhagen, they were politely received by Poulsen, who pleased Jensen by now addressing him as "Mr." in acknowledgment of the enhanced status he had gained through his experience in America. The Danish inventor, however, refused to participate in the scheme proposed by the San Francisco syndicate.

By this time, perhaps because of Pridham's influence, Jensen had decided to return to the United States and become an American citizen. After visiting his mother, he and Pridham headed back by way of London, sailing from there to New York City on the *Lusitania*. As the ship came through the Narrows and Jensen gazed at the Statue of Liberty, a "new and strong feeling of patriotism" swept over him. "I knew this country would be mine, forever," he later declared.

Returning to California, Jensen and Pridham were introduced by Coburn to Richard O'Connor, a San Francisco soap and candle manufacturer with powerful connections in local and state politics. Impressing Jensen as "the ideal type of politician in the best sense of the word," he was a strong backer of Hiram Johnson, who became governor of California in 1910 on a progressive reform ticket. O'Connor and his business associates had been involved in the abortive plan to acquire Poulsen's world patent rights, and were still eager to invest in radio despite the failure of the recent mission to Copenhagen. With their support, an enterprise called the Commercial Wireless and Development Company (CWDC) was organized, with O'Connor as president and treasurer. As the firm's engineers, Jensen and Pridham were to conduct research and experimentation in radio and take out patents on anything they might invent. On February 22, 1911, they moved to Napa, about thirty-five miles northwest of San Francisco, choosing the community for its relative seclusion. Acquiring "a small bungalow and a good sized lot," they assembled some equipment and commenced work. Within a short time Jensen met and married a Californian, Vivian Steves. By 1914 they had a daughter, Jean, and a son, Karl. Two more children, Patricia and Marian, followed in 1917 and 1925.

The project in which Jensen and Pridham became engaged in their Napa laboratory epitomizes the process of technological transfer that took place between key industries—in this case involving the telegraph, the telephone, and radio—as the development of continuous-wave technology gained momentum. Joined by Carl Albertus, who signed on as a mechanic, Jensen and Pridham became intrigued by the photographic apparatus that Poulsen had devised for recording high-speed transmissions of dots and dashes from his tikker. Poulsen had tried to sell rights to this device to Elwell and his American backers in 1908, but they had not been impressed by it. Struck by the rapidity with which the thin wire in the tikker responded to signals as it moved back and forth between the poles of a magnet, Jensen and Pridham thought it might respond just as rapidly to sound waves and furnish the basis for a telephone receiver. Substituting a somewhat heavier wire for the thin one, they connected it to a diaphragm with a matchstick and integrated it into a telephone circuit with a microphone speaker. Testing the new apparatus, they were gratified to hear that the receiver reproduced human speech "with exceptional strength and clarity."

After making a careful search of the scientific and technical literature at their disposal, Jensen and Pridham mistakenly concluded that they had stumbled upon a new method of sound reproduction. Applying for a patent covering what they described as the "electro-dynamic principle," they improved their new invention by suspending a coil of copper wire between stronger electromagnets and experimenting with a variety of diaphragms and soundboxes. Believing that they had invented a telephone that performed better than any then in use, they received a setback when their patent application was denied; the principle they had thought to be original had in fact been anticipated by a number of scientists and inventors in Germany, Great Britain, and the United States. Because nothing had ever been done to exploit it commercially, they did manage to win a subsequent patent protecting only their specific device. Nevertheless, it was still too heavy and expensive to compete with existing telephone receivers, and they were unable to interest the American Telephone and Telegraph Company (AT&T) in purchasing the rights to it despite a trip they made to New York to talk to officials at AT&T's Western Electric Laboratories.

Bitterly disappointed, Jensen and Pridham talked about ending their partnership on the way back to California. Having invested approximately $30,000 in their work, most of the firm's backers were also ready to pull out, but O'Connor was unwavering in his support and the venture stayed precariously alive. At this critical moment, Ray Galbreath, an uncle of Vivian Jensen, visited the Napa laboratory and suggested a new approach. Referring to a colorful local figure who made announcements with a megaphone at baseball games in the San Francisco area, Galbreath suggested that the failed receiver might serve as a new means of making public announcements. "If you can make it talk a little louder and put a horn on it, like Foghorn Murphy's," he said, "and if you put enough of them around a ball park, maybe we can understand what is being said a little better, and maybe they don't need Foghorn Murphy any more." This was a far cry from the lucrative telephone markets Jensen and Pridham had in mind, but there was also the prospect that the type of device Galbreath had suggested could be used in railroad stations to announce train departures and arrivals. Not knowing what else to do, they decided to proceed.

Taking a gooseneck horn from an old Edison phonograph, Pridham and Jensen designed a fitting to connect it with their receiver. Clustering six powerful microphones acquired during their European trip in 1910, they improvised an improved transmitter. To their original circuitry they added a transformer and a twelve-volt storage battery. Without realizing it, they had created a sound system with a potential output approaching twenty-five watts, thousands of times greater than anything in existence. As Jensen later recalled, connecting the components resulted in "a screaming howling noise which was ear-splitting and terrifying," but they recognized this as a feedback phenomenon that could be corrected if the transmitter and the receiver were placed far enough apart. Running a line to the roof of the bungalow, they attached the horn and receiver to the chimney, pointing northwest toward open countryside. When Pridham spoke through the microphone cluster inside the house, it sounded to Jensen "like a voice not of this earth. Had I closed my eyes it would have been easy to imagine that a supernatural colossus was shouting up the chimney."

Sprinting across the open fields with Albertus struggling unsuccessfully to keep up, Jensen found that by listening intently he could understand what Pridham was saying a mile away. When he returned, Pridham bicycled to the same spot and listened for himself, after which the excited associates called O'Connor in San Francisco. Incredulous at the news, O'Connor insisted that it be repeated several times. Satisfied at last that it was true, he brought a group of stockholders to Napa the following day, accompanied by a lawyer. The successful demonstration that followed produced assurances "that finances would no longer be a problem." Casting about for a name to give their brainchild, Jensen and Pridham considered such alternatives as "Stentor" and "Telemegaphone," but finally settled on "Magnavox," a term meaning "great voice" in Latin. Ultimately it became universally known as the "loudspeaker," a name Jensen had already considered but rejected because he found it unappealing.

At this point, Jensen and Pridham had gone through a process that historians of technology would subsequently call "invention," in which a new principle or device appears in rudimentary form. They now entered upon a stage that scholars would later characterize as "development," in which modifications are made to improve a new technology to prepare it for commercial application in a final stage known as "innovation." Through 1915 Jensen and Pridham gradually improved their invention. Some of the changes they made merely affected its appearance, giving it the distinctive shape that would become characteristic of loudspeakers in the future. Others were more substantive, such as substituting new American-made microphone buttons for the previous European ones in order to enhance transmission. In the process, they created the prototype of the electrical phonograph. Placing the loudspeaker in a cabinet, they mounted a turntable and microphone on top and designed a volume control that could be turned up or down by manipulating a knob. A patent was quickly obtained, the only remotely comparable device being a box within which a telephone receiver served as a burglar alarm by emitting a whistling sound.

During the course of their work, Jensen and Pridham treated the townspeople of Napa to occasional concerts of recorded music from their rooftop. A local resident was startled one evening to hear an "extraordinary voice" coming from three-quarters of a mile away; it turned out to be that of the famous contralto and mezzo-soprano, Ernestine Schumann-Heink, singing Christmas

Edwin S. Pridham (left) and Peter L. Jensen in their laboratory in Napa, California, in 1915.

music. Ultimately the sound of the loudspeaker could be heard within a radius of seven miles.

Fearing that premature exposure of their system would stimulate efforts to copy it, Jensen and Pridham refrained from exhibiting it at the 1915 Panama Pacific World's Exposition in San Francisco. It was obvious that a formal public demonstration could not be indefinitely postponed, however, and on December 10, 1915, such an event took place before an invited group in the stadium at Golden Gate Park. Conditions were far from ideal; the wind was blowing at twenty miles per hour, there was a drizzle of rain, and some schoolboys played football on the field as the test took place. Nevertheless, reporter Edgar ("Scoop") Gleason published an enthusiastic article about the demonstration the following day in the San Francisco *Bulletin*, commenting in glowing terms about "the slender tone of a single violin heard one mile away," the sound of operatic soprano Luisa Tetrazzini's voice reverberating throughout the stadium, and "a piano solo resembling the chimes of Westminster Abbey played by the Colossus of Rhodes." Although the loudspeaker was placed at a considerable distance from the listeners, the noise of the football game was inaudible as the music of a Hawaiian string quartet resounded "like the harps of the cyclops."

Two weeks later, on Christmas Eve, a much more elaborate airing of the system took place when a crowd estimated at 100,000 people gathered in the plaza in front of San Francisco's new city hall and listened to recorded music that sounded as if it had been "uttered by a giant" as part of an annual holiday celebration sponsored by the *Bulletin*. Still concerned about keeping the nature of their apparatus as secret as possible, Jensen and Pridham operated it from a balcony, the loudspeaker being concealed behind an American flag.

On December 30, the first indoor demonstration of the loudspeaker took place in San Francisco's new Civic Auditorium, a 12,000-seat facility that had recently been given to the city by the Panama Pacific Exposition Company. A speech by Governor Johnson was to be transmitted to the hall from his mansion on Green Street, two miles away, and special wiring arrangements were made to facilitate this. Jensen, stationed in the hall while Pridham supervised transmitting equipment located in the mansion, was understandably nervous as the opening presentations got underway. He agonized as the architect who had designed the building, unaccustomed to the use of a loudspeaker, became alternately inaudible and overpowering as he waved the microphone back and forth while pointing to various features of the auditorium. After this ordeal was over, the young Dane precipitated a minor scuffle trying to disperse a choir that had massed in front of the horn from which the Governor's voice was to emanate. Despite such problems, and the fact that Johnson's speech could be heard much better in the galleries than was possible on the main floor because of the placement of the choir, the address received a loud ovation and newspaper coverage the following day was enthusiastic.

However successful, these demonstrations exposed remaining difficulties, particularly the need for a means of electrically amplifying weak signals, such as those coming from Governor Johnson's mansion, and using it to drive their loudspeaker. This was ultimately solved through the use of Lee de Forest's audion tube, which had been used to spectacular effect by AT&T in 1915 in radio transmissions between Paris, Virginia, and Honolulu. Precisely how the rights to the audion were obtained is not clear, but this may have resulted from a private agreement with de Forest, who had sold his patent rights to AT&T but retained permission to sell equipment to "amateurs," a term that de Forest construed quite liberally.

Having crossed this technological hurdle, the two inventors still faced other problems. The San Francisco demonstrations did not lead immediately to marketing opportunities, causing Jensen and Pridham to suspect that public speakers looked upon the loudspeaker as an aspersion on their histrionic abilities. As the two inventors and their financial backers pondered an apparent lack of interest among prospective users of public address systems, they decided that adapting the loudspeaker for use in an electronically amplified phonograph might have better market appeal. This, however, would require additional capital and a familiarity with the phonograph industry that neither Jensen and Pridham nor their backers possessed. Although efforts to interest the Columbia and Victor phonograph companies in their device proved fruitless, a chance encounter with George Cook Sweet, a naval officer who visited their Napa laboratory and was impressed with the amplifying power of their equipment, led to a crucial contact with Frank

M. Steers, president of the Sonora Phonograph Corporation in San Francisco. Late in 1916 Jensen and Pridham sold the Napa laboratory and moved to that city. On August 3, 1917, the Commercial Wireless and Development Company was consolidated with Sonora and renamed the Magnavox Company. Although O'Connor remained a director in the new enterprise, Steers became its chief executive, while Jensen and Pridham were jointly designated "Chief Engineer."

By this time the United States had entered World War I, delaying the phonograph gambit and leading Magnavox to confine itself to military applications. This opened up new prospects for the loudspeaker. In cooperation with the U.S. Navy, the company took part in an unsuccessful attempt to use the new device to facilitate communication between a ground station and a plane flying overhead. Much more successful was a project in which, by stripping a microphone of its casing and mouthpiece and exposing its diaphragm to the open air, Jensen and Pridham developed an anti-noise transmitter that made the human voice audible over the roar of an airplane engine. Using earphones encased in rubber inside the helmets of crew members, this permitted communication within an airplane during flight. Widely used during the remainder of the war, the system also saw service in the Curtiss NC-4 flying boats used in the U.S. Navy's pioneering transatlantic flight from Newfoundland to Lisbon in 1919. Jensen and Pridham also developed a public-address system for use on naval vessels, particularly amid the noise of engine rooms or during storms.

After the return of peace in 1918 the Magnavox Company was kept busy making watertight telephones for use on ships and little was done to exploit the promise inherent in the loudspeaker. This suddenly changed, however, when President Woodrow Wilson came to California during his nationwide campaign to promote American ratification of the Treaty of Versailles and membership in the League of Nations. Forbidden by his physicians to speak out-of-doors because of his failing health, it appeared that he would not be able to make a scheduled address in San Diego. Seeking to forestall its cancellation, a group of citizens approached the Magnavox Company about the possibility of installing a loudspeaker system in the municipal stadium so that Wilson could speak from a glass-enclosed booth. Magnavox agreed, and appropriate preparations were made. Because Jensen was in the East

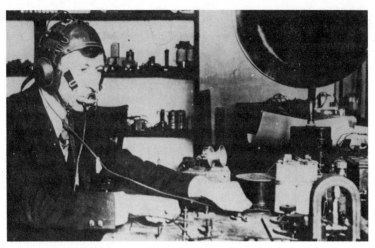

Jensen in his laboratory in 1917. The noise-neutralizing microphone he is wearing was invented and patented by Jensen and Pridham.

on a business trip and could not attend the event, which took place on August 19, 1919, Pridham was put in charge of what turned out to be a briefly harrowing assignment. Playing recorded music to entertain the crowd of 50,000 while Wilson's motorcade was taking part in a parade en route to the stadium, Pridham was horrified when the system went dead at the very moment the President's automobile came into view. As smoke began coming out of the amplifier, Pridham desperately removed the nearest tube from its socket and replaced it with a new one. Providentially, it worked; somehow he had gone directly to the source of the short circuit that had caused the difficulty. The following day, articles in such newspapers as the Chicago *Tribune* reported how Wilson, speaking in a normal tone of voice, was heard distinctly through two Magnavox speakers "with as much ease as if he were talking in a small banquet room."

Practically overnight, the loudspeaker now received rapid national and international attention and acceptance. Loudspeakers were used at both the Democratic and Republican national conventions in 1920, and both presidential candidates, James M. Cox and Warren G. Harding, used them extensively in their campaigns. A loudspeaking system was also used at Harding's inauguration in Washington on March 4, 1921. Suddenly the market for loudspeakers appeared limitless, forcing Magnavox to reassess its future. Company leaders feared that they could not meet mounting demand with the capital at their disposal and also faced the inevitable consequences of a struggle with AT&T, which along with General Electric had a commanding position in vacuum-tube technology. The California firm's West Coast location posed an additional problem, making it difficult to compete in key Eastern markets. Magnavox therefore yielded most of the market for public address systems to AT&T. Reverting to a strategy already formulated in 1917, from 1922 onward it devoted itself chiefly to the booming radio and phonograph industry, producing a succession of improved horn speakers, receivers, and vacuum tubes.

Now famous and with his financial position secure, Jensen went to Europe in 1922 to study developments in loudspeaker technology taking place there, to develop new markets for Magnavox products in England, and to visit his native land, where his mother still lived. Spending the summer in London, he staged the first demonstration of the loudspeaker ever to take place in that city in connection with a prize fight between George Carpenter and Ted ("Kid") Lewis, which resulted in a scoop for his sponsor, the *Daily Mail*. He also met the father of radio, Guglielmo Marconi, for the first time and chatted with him about Valdemar Poulsen and the early days of wireless communication. Proceeding to Denmark, he demonstrated his loudspeaker system to a throng gathered in City Hall Square in Copenhagen, creating a sensation when he amplified a long-distance telephone conversation with a correspondent in Stockholm so that the entire crowd could hear it.

Finally, Jensen went back to Falster, where his mother was taken aback by the unexpectedly resounding kiss with which he greeted her. "I had learned to show my affections in America," he later explained. "I hear you are famous," she responded; "now I only hope it will last!" During his visit, he tramped around the island, reminiscing about old times. Gazing at the Strait of Groensund from a cliff overlooking the Baltic Sea, he was overcome by intense feelings of love for his native land. It made him realize that, although he had "striven to throw off every vestige of a Dane in my effort to become a better citizen of my adopted country," he had become both "a better American and a better Dane than I had been before."

Returning to America, Jensen ultimately broke with other Magnavox executives over policy matters. Though Pridham remained with the company, Jensen left it in 1925. In 1927 he founded the Jensen Radio Manufacturing Company; originally headquartered in Oakland, California, it was soon moved to Chicago, where, with the aid of engineer Hugh Knowles, Jensen worked intensively to eliminate distortion and improve fidelity in sound reproduction. In this new phase of his career Jensen's marriage to his wife Vivian ended in divorce; in 1929 he married Malvena Opliger, by whom he had one child. In 1943, disputes with financial backers led once more to his resignation from a firm of his own creation, and he subsequently founded Jensen Industries to manufacture phonograph needles. He also served as chief consultant to the U.S. War Production Board's radio and radar division in World War II.

Jensen's old age brought him many honors, perhaps the most important being the conferring of knighthood by the King of Denmark in 1956. He also received awards from the American Institute of Radio Engineers and the Audio Engineering Society. A man of many interests, he studied philosophy, played the violin, enjoyed light opera, and took part avidly in various sports, particularly tennis. An inveterate smoker, he died of lung cancer in Western Springs, Illinois, on October 26, 1961. His friendship with Pridham, who died in 1963, had continued unabated throughout the years.

Jensen's career was aptly summarized by an obituary in the *Journal of the Audio Engineering Society*: "Of the many men who distinguished themselves in the early years of radio engineering, Peter Jensen was undoubtedly one of the most creative and productive. His inventions, experiments and sound systems were the heralds of today's high fidelity industry." In a broader perspective, Jensen's career provided a textbook case of how an important innovation may proceed out of a rich background of technological effort, in this case the worldwide pattern of engineering activity that accompanied the emergence of radio in the late nineteenth and early twentieth centuries, and, by going through characteristic stages of invention and development, yield a product that, through commercial exploitation, can become an important part of everyday life. Jensen's lifetime spanned a period in which a long list of innovations including the telephone, the phonograph, motion pictures, radio, and television transformed human communications; the loudspeaker figured importantly in enhancing the power and impact of every one of them. In human terms, the life of this

Danish-American is a case study of the path followed by a legion of immigrant inventors, including such persons as Alexander Graham Bell, Michael Pupin, Nikola Tesla, Charles P. Steinmetz, and Vladimir Zworykin, who found the United States to be truly a land of opportunity and contributed to a process of technological transformation that continues to unfold in the late twentieth century.

Bibliograpical Note

Jensen's life can be traced in his autobiography, *En verdenskendt Dansker, Jensen, hottalerens opfinder: En selvbiografi* (Copenhagen: C. Erichsen, 1948), subsequently published in English as *The Great Voice* (Richardson, TX: The Havilah Press, 1975). I have drawn heavily upon both versions, and wish to thank Professor Ole Krogh, formerly of Auburn University, for his assistance in working with the Danish original. Supplementary information on Jensen's life is available in standard Danish biographical dictionaries, including *Dansk Biografisk Leksikon* (1937) and *Kraks blaa bog* (1949, 1961). For a brief synopsis of his career, see my sketch in John A. Garraty, ed., *Dictionary of American Biography* (New York: Charles Scribner's Sons, 1981), Supplement VII (1961–1965), s.v. "Jensen, Peter Laurits."

For authoritative analysis of the background of radio throughout Jensen's career, see Hugh G. J. Aitken, *Syntony and Spark: The Origins of Radio* (Princeton: Princeton University Press, 1976) and *The Continuous Wave: Technology and American Radio, 1900–1932* (Princeton: Princeton University Press, 1985). I greatly appreciate Aitken's help and advice as this essay went through successive drafts; thanks are also due to Bernard Finn and Elliot Sivowitch of the National Museum of American History, the Scandanavian Section of the Library of Congress, and Boyd Childress of the Ralph Brown Draughon Library of Auburn University for help at various stages of my research on Jensen's career. On the invention-development-innovation model of technological change, see Peter F. Drucker, "Technological Trends in the Twentieth Century," in Melvin Kranzberg and Carroll W. Pursell, *Technology in Western Civilization*, II (New York: Oxford University Press, 1967); Devendra Sahal, *Patterns of Technological Innovation* (Reading, MA: Addison-Wesley, 1981); and John M. Staudenmaier, *Technology's Storytellers: Reweaving the Human Fabric* (Cambridge, MA: MIT Press, 1985), usefully supplemented by a series of articles by Thomas P. Hughes, Lynwood Bryant, Thomas M. Smith, Richard G. Hewlett, and Charles Susskind in a special section on "The Development Phase of Technological Change" in *Technology and Culture*, XVII (July, 1976). On the "shop culture" model into which Jensen's training under Poulsen fits, see Monte A. Calvert, *The Mechanical Engineer in America, 1830–1910: Professional Cultures in Conflict* (Baltimore: Johns Hopkins Press, 1967). On the development of the American broadcasting industry during the period in which Jensen lived, see particularly Erik Barnouw, *A History of Broadcasting in the United States*, 3 vols. (New York: Oxford University Press, 1966–1970), and Susan J. Douglas, *Inventing American Broadcasting, 1899–1922* (Baltimore and London: Johns Hopkins University Press, 1987).

For additional perspectives on Jensen's career, see Glenn D. Kittler, "Forgotten Man of Sound," *Coronet*, May, 1954; Jane Morgan, *Electronics in the West: The First Fifty Years* (Palo Alto: National Press Books, 1967); Lillian C. White, *Pioneer and Patriot: George Cook Sweet, Commander, U.S.N.* (Delray Beach, FL: Southern Publishing Company, 1963); and Robert Lozier, "Twenty Years of the Magnavox Story: 1911–1931," an undated reprint from the *Bulletin* of the Antique Wireless Association supplied to me by Aitken. An obituary of Jensen is in the *Journal of the Audio Engineering Society*, January, 1961.

18

CHARLES A. LINDBERGH: HIS FLIGHT AND THE AMERICAN IDEAL*

John William Ward

On Friday, May 20, 1927, at 7:52 a.m., Charles A. Lindbergh took off in a silver-winged monoplane and flew from the United States to France. With this flight Lindbergh became the first man to fly alone across the Atlantic Ocean. The log of flight 33 of "The Spirit of St. Louis" reads: "Roosevelt Field, Long Island, New York, to Le Bourget Aerodrome, Paris, France. 33 hrs. 30 min." Thus was the fact of Lindbergh's achievement easily put down. But the meaning of Lindbergh's flight lay hidden in the next sentence of the log: "(Fuselage fabric badly torn by souvenir hunters.)"

When Lindbergh landed at Le Bourget he is supposed to have said, "Well, we've done it." A contemporary writer asked "Did what?" Lindbergh "had no idea of what he had done. He thought he had simply flown from New York to Paris. What he had really done was something far greater. He had fired the imagination of mankind." From the moment of Lindbergh's flight people recognized that something more was involved than the mere fact of the

---

*This essay, originally titled "The Meaning of Lindbergh's Flight," is reprinted by the courtesy of *The American Quarterly* from its Spring, 1958 issue.

physical leap from New York to Paris. "Lindbergh," wrote John Erskine, "served as a metaphor." But what the metaphor stood for was not easy to say. The *New York Times* remarked then that "there has been no complete and satisfactory explanation of the enthusiasm and acclaim for Captain Lindbergh." Looking back on the celebration of Lindbergh, one can see now that the American people were trying to understand Lindbergh's flight, to grasp its meaning, and through it, perhaps, to grasp the meaning of their own experience. Was the flight the achievement of a heroic, solitary, unaided individual? Or did the flight represent the triumph of the machine, the success of an industrially organized society? These questions were central to the meaning of Lindbergh's flight. They were also central to the lives of the people who made Lindbergh their hero.

The flight demanded attention in its own right, of course, quite apart from whatever significance it might have. Lindbergh's story had all the makings of great drama. Since 1919 there had been a standing prize of $25,000 to be awarded to the first aviator who could cross the Atlantic in either direction between the United States and France in a heavier-than-aircraft. In the spring of 1927 there promised to be what the *New York Times* called "the most spectacular race ever held—3,600 miles over the open sea to Paris." The scene was dominated by veteran pilots. On the European side were the French aces, Nungesser and Coli; on the American side, Commander Richard E. Byrd, in a big tri-motored Fokker monoplane, led a group of contestants. Besides Byrd, who had already flown over the North Pole, there were Commander Davis, flying a ship named in honor of the American Legion which had put up $100,000 to finance his attempt, Clarence Chamberlin, who had already set a world's endurance record of more than fifty-one hours in the air in a Bellanca tri-motored plane, and Captain René Fonck, the French war ace, who had come to America to fly a Sikorsky aircraft. The hero was unheard of and unknown. He was on the West Coast supervising the construction of a single-engined plane to cost only ten thousand dollars.

Then fate played its part. It seemed impossible that Lindbergh could get his plane built and east to New York in time to challenge his better equipped and more famous rivals. But in quick succession a series of disasters cleared his path. On April 16 Commander Byrd's "America" crashed on its test flight, crushing the leg of

Floyd Bennett who was one of the crew and injuring Byrd's hand and wrist. On April 24 Clarence Chamberlin cracked up in his Bellanca, not seriously, but enough to delay his plans. Then on April 26 Commander Davis and his co-pilot lost their lives as the "American Legion" crashed on its final test flight. In ten days, accidents had stopped all of Lindbergh's American rivals. Nungesser and Coli, however, took off in their romantically named ship, "The White Bird," from Le Bourget on May 8. The world waited and Lindbergh, still on the West Coast, decided to try to fly the Pacific, But Nungesser and Coli were never seen again. As rumors filled the newspapers, as reports came in that the "White Bird" was seen over Newfoundland, over Boston, over the Atlantic, it soon became apparent that Nungesser and Coli had failed, dropping to their death in some unknown grave. Disaster had touched every ship entered in the trans-Atlantic race.

Now, with the stage cleared, Lindbergh entered. He swooped across the continent in two great strides, landing only at St. Louis. The first leg of his flight established a new distance record but all eyes were on the Atlantic and the feat received little notice. Curiously, the first time Lindbergh appeared in the headlines of the New York papers was Friday, the 13th. By this time Byrd and Chamberlin were ready once again but the weather had closed in and kept all planes on the ground. Then, after a week of fretful waiting, on the night of May 19 on the way into New York to see "Rio Rita," Lindbergh received a report that the weather was breaking over the ocean. He hurried back to Roosevelt Field to haul his plane out onto a wet, dripping runway. After mechanics painfully loaded the plane's gas by hand, the wind shifted, as fate played its last trick. A muddy runway and an adverse wind. Whatever the elements, whatever the fates, the decisive act is the hero's, and Lindbergh made his choice. Providing a chorus to the action, the *Herald Tribune* reported that Lindbergh lifted the overloaded plane into the sky "by his indomitable will alone."

The parabola of the action was as clean as the arc of Lindbergh's flight. The drama should have ended with the landing of "The Spirit of St. Louis" at Le Bourget. That is where Lindbergh wanted it to end. In *"WE,"* written immediately after the flight, and in *The Spirit of St. Louis*, written 26 years later, Lindbergh chose to end his accounts there. But the flight turned out to be only the first act in the part Lindbergh was to play.

Lindbergh was so innocent of his future that on his flight he carried letters of introduction. The hysterical response, first of the French and then of his own countrymen, had been no part of his

Acclaimed as the "Lone Eagle," Charles A. Lindbergh poses beside his monoplane, "The Spirit of St. Louis."

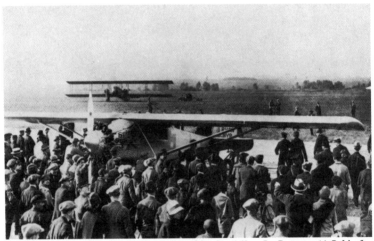

A crowd gathers in front of "The Spirit of St. Louis" at Le Bourget Airfield after Lindbergh's solo nonstop flight on May 20–21, 1927.

Source: UPI

214

careful plans. In *"WE,"* after Lindbergh's narrative of the flight, the publisher wrote: "When Lindbergh came to tell the story of his welcome at Paris, London, Brussels, Washington, New York, and St. Louis, he found himself up against a tougher problem than flying the Atlantic." So another writer completed the account in the third person. He suggested that "the reason Lindbergh's story is different is that when his plane came to a halt on Le Bourget field that black night in Paris, Lindbergh the man kept on going. The phenomenon of Lindbergh took its start with his flight across the ocean; but in its entirety it was almost as distinct from that flight as though he had never flown at all."

Lindbergh's private life ended with his flight to Paris. The drama was no longer his, it was the public's. "The outburst of unanimous acclaim was at once personal and symbolic," said the *American Review of Reviews*. From the moment of success there were two Lindberghs, the private Lindbergh and the public Lindbergh. The latter was the construction of the imagination of Lindbergh's time, fastened on to an unwilling person. The tragedy of Lindbergh's career is that he could never accept the role assigned him. He always believed he might keep his two lives separate. But from the moment he landed at Le Bourget, Lindbergh became, as the *New Republic* noted, *"ours* .... He is no longer permitted to be himself. He is US personified. He is the United States." Ambassador Herrick introduced Lindbergh to the French, saying, "This young man from out of the West brings you better than anything else the spirit of America," and wired to President Coolidge, "Had we searched all America we could not have found a better type than young Lindbergh to represent the spirit and high purpose of our people." This was Lindbergh's fate, to be a type. A writer in the *North American Review* felt that Lindbergh represented "the dominant American character," he "images the best" about the United States. And an ecstatic female in the *American Magazine,* who began by saying that Lindbergh "is a sort of symbol .... He is the dream that is in our hearts," concluded that the American public responded so wildly to Lindbergh because of "the thrill of possessing, in him, our dream of what *we* really and truly want to be." The act of possession was so complete that articles since have attempted to discover the "real" Lindbergh, that enigmatic and taciturn figure behind the public mask. But it is no less difficult to discern the features of the public Lindbergh, that symbolic figure who presented to the imagination of his time all the yearnings and buried desires of its dream for itself.

Lindbergh's flight came at the end of a decade marked by social and political corruption and by a sense of moral loss. The heady idealism of the First World War had been succeeded by a deep cynicism as to the war's real purpose. The naïve belief that virtue could be legislated was violated by the vast discrepancy between the law and the social habits of prohibition. A philosophy of relativism had become the uneasy rationale of a nation which had formerly believed in moral absolutes. The newspapers agreed that Lindbergh's chief worth was his spiritual and moral value. His story was held to be "in striking contrast with the sordid unhallowed themes that have for months steeped the imaginations and thinking of the people." Or, as another had it, "there is good reason why people should hail Lindbergh and give him honor. He stands out in a grubby world as an inspiration."

Lindbergh gave the American people a glimpse of what they liked to think themselves to be at a time when they feared they had deserted their own vision of themselves. The grubbiness of the 1920's had a good deal to do with the shining quality of Lindbergh's success, especially when one remembers that Lindbergh's flight was not as unexampled as our national memory would have it. The Atlantic was not unconquered when Lindbergh flew. A British dirigible had twice crossed the Atlantic before 1919 and on May 8 of that year three naval seaplanes left Rockaway, New York, and one, the NC-4 manned by a crew of five, got through to Plymouth, England. A month later, Captain John Alcock, an Englishman, with Arthur W. Browne, an American, flew the first heavier-than-air land plane across the Atlantic nonstop, from Newfoundland to Ireland, to win twice the money Lindbergh did, a prize of $50,000 offered by the London *Daily Mail*. Alcock's and Browne's misfortune was to land in a soft and somnolent Irish peat bog instead of before the cheering thousands of London or Paris. Or perhaps they should have flown in 1927.

The wild medley of public acclaim and the homeric strivings of editors make one realize that the response to Lindbergh involved a mass ritual in which America celebrated itself more than it celebrated Lindbergh. Lindbergh's flight was the occasion of a public act of regeneration in which the nation momentarily rededicated itself to something, the loss of which was keenly felt. It was said

New York welcomes Lindbergh with a triumphal ticker-tape parade on June 14, 1927.

Source: UPI

again and again that "Lindy" taught America "to lift its eyes up to Heaven." Heywood Broun, in his column in the *New York World*, wrote that this "tall young man raised up and let us see the potentialities of the human spirit." Broun felt that the flight proved that, though "we are small and fragile," it "isn't true that there is no health in us." Lindbergh's flight provided the moment, but the meaning of the flight is to be found in the deep and pervasive need for renewal which the flight brought to the surface of public feeling. When Lindbergh appeared at the nation's capital, the *Washington Post* observed, "He was given that frenzied acclaim which comes from the depths of the people." In New York, where 4,000,000 people saw him, a reporter wrote that the dense and vociferous crowds were swept, as Lindbergh passed, "with an emotion tense and inflammable." The *Literary Digest* suggested that the answer to the hero worship of Lindbergh would "throw an interesting light on the psychology of our times and of the American people."

The *Nation* noted about Lindbergh that "there was something lyric as well as heroic about the apparition of this young Lochinvar who suddenly came out of the West and who flew all unarmed and all alone. It is the kind of stuff which the ancient Greeks would have worked into a myth and the medieval Scots into a border ballad . . . . But what we have in the case of Lindbergh is an actual, an heroic and an exhaustively exposed experience which exists by suggestion in the form of poetry." The *Nation* quickly qualified its statement by observing that reporters were as far as possible from being poets and concluded that the discrepancy between the fact and the celebration of it was not poetry, perhaps, but "magic on a vast scale." Yet the *Nation* might have clung to its insight that the public meaning of Lindbergh's flight was somehow poetic. The vast publicity about Lindbergh corresponds in one vital particular with the poetic vision. Poetry, said William Butler Yeats, contains opposites; so did Lindbergh. Lindbergh did not mean one thing, he meant many things. The image of itself which America contemplated in the public person of Lindbergh was full of conflict; it was, in a word, dramatic.

To heighten the drama, Lindbergh did it alone. He was the "lone eagle" and a full exploration of that fact takes one deep into the emotional meaning of his success. Not only the *Nation* found Sir Walter Scott's lines on Lochinvar appropriate: "he rode all unarmed and he rode all alone." Newspapers and magazines were deluged with amateur poems that vindicated one rhymester's wry comment, "Go conquer the perils / That lurk in the skies - - / And you'll get bum poems / Right up to your eyes." The *New York Times*, that alone received more than two hundred poems, observed in trying to summarize the poetic deluge that "the fact that he flew alone made the strongest impression." Another favorite tribute was Kipling's "The Winners," with its refrain, "He travels the fastest who travels alone." The others who had conquered the Atlantic and those like Byrd and Chamberlin who were trying at the same time were not traveling alone and they hardly rode unarmed. Other than Lindbergh, all the contestants in the trans-Atlantic race had unlimited backing, access to the best planes, and all were working in teams, carrying at least one co-pilot to share the long burden of flying the plane. So a writer in the New York *Sun,* in a poem called "The Flying Fool," a nickname that Lindbergh despised, celebrated Lindbergh's flight: "... no kingly plane for him;/ No endless data, comrades, moneyed chums;/ No boards, no councils, no directors grim— / He plans ALONE... and takes luck as it comes."

Upon second thought, it must seem strange that the long distance flight of an airplane, the achievement of a highly advanced and organized technology should be the occasion for hymns of praise to the solitary unaided man. Yet the National Geographic Society, when it presented a medal to Lindbergh, wrote on the presentation scroll, "Courage, when it goes alone, has ever caught men's imaginations," and compared Lindbergh to Robinson Crusoe and the trailmakers in our own West. But Lindbergh and Robinson Crusoe, the one in his helmet and fur-lined flying coat and the other in his wild goatskins, do not easily coexist. Even if Robinson Crusoe did have a tidy capital investment in the form of a well-stocked shipwreck, he still did not have a ten-thousand-dollar machine under him.

Lindbergh, in nearly every remark about his flight and in his own writings about it, resisted the tendency to exploit the flight as the achievement of an individual. He never said "I," he always said "We." The plane was not to go unrecognized. Nevertheless, there persisted a tendency to seize upon the flight as a way of celebrating the self-sufficient individual, so that among many others an Ohio newspaper could describe Lindbergh as this "self-contained, self-reliant, courageous young man [who] ranks among the great pioneers of history." The strategy here was a common one, to make Lindbergh a "pioneer" and thus to link him with a long and vital tradition of individualism in the American experience. Colonel Theodore Roosevelt, himself the son of a famous exponent of self-reliance, said to reporters at his home in Oyster Bay that

"Captain Lindbergh personifies the daring of youth. Daniel Boone, David Crocket [sic], and men of that type played a lone hand and made America. Lindbergh is their lineal descendant." In *Outlook* magazine, immediately below an enthusiastic endorsement of Lindbergh's own remarks on the importance of his machine and his scientific instruments, there was the statement, "Charles Lindbergh is the heir of all that we like to think is best in America. He is of the stuff out of which have been made the pioneers that opened up the wilderness, first on the Atlantic coast, and then in our great West. His are the qualities which we, as a people, must nourish." It is in this mood that one suspects it was important that Lindbergh came out of the West and rode all alone.

Another common metaphor in the attempt to place Lindbergh's exploit was to say that he had opened a new "frontier." To speak of the air as a "frontier" was to invoke an interpretation of the meaning of American history which had sources deep in American experience, but the frontier of the airplane is hardly the frontier of the trailmakers of the old West. Rather than an escape into the self-sufficient simplicity of the American past, the machine which made Lindbergh's flight possible represented an advance into a complex industrial present. The difficulty lay in using an instance of modern life to celebrate the virtues of the past, to use an extreme development of an urban industrial society to insist upon the significance of the frontier in American life.

A little more than a month after Lindbergh's flight, Joseph K. Hart in *Survey* magazine reached back to Walt Whitman's poem for the title of an article on Lindbergh: "O Pioneer." A school had made Lindbergh an honorary alumnus but Hart protested there was little available evidence "that he was educated in *schools.*" "We must look elsewhere for our explanation," Hart wrote and he looked to the experience of Lindbergh's youth when "everything that he ever did . . . he did by himself. He lived more to himself than most boys." And, of course, Lindbergh lived to himself in the only place conceivably possible, in the world of nature, on a Minnesota farm. "There he developed in the companionship of woods and fields, animals and machines, his audaciously natural and simple personality," The word, "machines," jars as it intrudes into Hart's idyllic pastoral landscape and betrays Hart's difficulty in relating the setting of nature upon which he wishes to insist with the fact that its product spent his whole life tinkering with machines, from motorcycles to airplanes. But except for that one word, Hart proceeds in uncritical nostalgia to show that "a lone trip across the Atlantic was not impossible for a boy who had grown up in the solitude of the woods and waters." If Lindbergh was "clear-headed, naif, untrained in the ways of cities," it was because he had "that 'natural simplicity' which Fenimore Cooper used to attribute to the pioneer hero of his Leatherstocking Tales." Hart rejected the notion that any student "bent to all the conformities" of formal training could have done what Lindbergh did. "Must we not admit," he asked, "that this pioneering urge remained to this audacious youth because he had never submitted completely to the repressions of the world and its jealous institutions?"

Only those who insist on reason will find it strange that Hart should use the industrial achievement of the airplane to reject the urban, institutionalized world of industrialism. Hart was dealing with something other than reason; he was dealing with the emotion evoked by Lindbergh's solitude. He recognized that people wished to call Lindbergh a "genius" because that "would release him from the ordinary rules of existence." That way, "we could rejoice with him in his triumph, and then go back to the contracted routines of our institutional ways [because] 99 percent of us must be content to be shaped and moulded by the routine ways and forms of the world to the routine tasks of life." It is in the word "must" that the pathos of this interpretation of the phenomenon of Lindbergh lies. The world had changed from the open society of the pioneer to the close-knit, interdependent world of a modern machine-oriented civilization. The institutions of a highly corporate industrial society existed as a constant reproach to a people who liked to believe that the meaning of its experience was embodied in the formless, independent life of the frontier. Like Thomas Jefferson who identified American virtue with nature and saw the city as a "great sore" on the public body, Hart concluded that "certainly, in the response that the world—especially the world of great cities—has made to the performance of this midwestern boy, we can read of the homesickness of the human soul, immured in city canyons and routine tasks, for the freer world of youth, for the open spaces of the pioneer, for the joy of battling with nature and clean storms once more on the frontiers of the earth."

221

The social actuality which made the adulation of Lindbergh possible had its own irony for the notion that America's strength lay in its simple uncomplicated beginnings. For the public response to Lindbergh to have reached the proportions it did, the world had by necessity to be the intricately developed world of modern mass communications. But more than irony was involved. Ultimately, the emotion attached to Lindbergh's flight involved no less than a whole theory about American history. By singling out the fact that Lindbergh rode alone, and by naming him a pioneer of the frontier, the public projected its sense that the source of America's strength lay somewhere in the past and that Lindbergh somehow meant that America must look backward in time to rediscover some lost virtue. The mood was nostalgic and American history was read as a decline, a decline measured in terms of America's advance into an urban, institutionalized way of life which made solitary achievement increasingly beyond the reach of 99 percent of the people. Because Lindbergh's ancestors were Norse, it was easy to call him a "Viking" and extend the emotion far into the past when all frontiers were open. He became the "Columbus" of another new world to conquer as well as the "Lochinvar" who rode all alone. But there was always the brute, irreducible fact that Lindbergh's exploit was a victory of the machine over the barriers of nature. If the only response to Lindbergh had been a retreat to the past, we would be involved with a mass cultural neurosis, the inability of America to accept reality, the reality of the world in which it lived. But there was another aspect, one in which the public celebrated the machine and the highly organized society of which it was a product. The response to Lindbergh reveals that the American people were deeply torn between conflicting interpretations of their own experience. By calling Lindbergh a pioneer, the people could read into American history the necessity of turning back to the frontier past. Yet the people could also read American history in terms of progress into the industrial future. They could do this by emphasizing the machine which was involved in Lindbergh's flight.

Lindbergh came back from Europe in an American man-of-war, the cruiser *Memphis*. It seems he had contemplated flying on, around the whole world perhaps, but less adventurous heads prevailed and dictated a surer mode of travel for so valuable a piece of public property. The *New Republic* protested against bringing

America's hero of romance home in a war ship. If he had returned on a great liner, that would have been one thing. "One's first trip on an oceanliner is a great adventure—the novelty of it, the many people of all kinds and conditions, floating for a week in a tiny compact world of their own." But to return on the *Memphis*, "to be put on a gray battleship with a collection of people all of the same stripe, in a kind of ship that has as much relation to the life of the seas as a Ford factory has! We might as well have put him in a pneumatic tube and shot him across the Atlantic." The interesting thing about the *New Republic's* protest against the unromantic, regimented life of a battleship is that the image it found appropriate was the Ford assembly line. It was this reaction against the discipline of a mechanized society that probably led to the nostalgic image of Lindbergh as a remnant of a past when romance was possible for the individual, when life held novelty and society was variegated rather than uniform. But what the Ford assembly line represents, a society committed to the path of full mechanization, was what lay behind Lindbergh's romantic success. A long piece in the Sunday *New York Times*, "Lindbergh Symbolizes the Genius of America," reminded its readers of the too obvious fact that "without an airplane he could not have flown at all." Lindbergh "is, indeed, the Icarus of the twentieth century; not himself an inventor of his own wings, but a son of that omnipotent Daedalus whose ingenuity has created the modern world." The point was that modern America was the creation of modern industry. Lindbergh "reveres his 'ship' as a noble expression of mechanical wisdom . . . . Yet in this reverence . . . Lindbergh is not an exception. What he means by the Spirit of St. Louis is really the spirit of America. The mechanical genius, which is discerned in Henry Ford as well as in Charles A. Lindbergh, is in the very atmosphere of [the] country." In contrast to a sentiment that feared the enforced discipline of the machine there existed an attitude of reverence for its power.

Lindbergh led the way in the celebration of the machine, not only implicitly by including his plane when he said "we," but by direct statement. In Paris he told newspapermen, "You fellows have not said enough about that wonderful motor." Rarely have two more taciturn figures confronted one another than when Lindbergh returned to Washington and Calvin Coolidge pinned the Distinguished Flying Cross on him, but in his brief remarks

Coolidge found room to express his particular delight that Lindbergh should have given equal credit to the airplane. "For we are proud," said the President, "that in every particular this silent partner represented American genius and industry. I am told that more than 100 separate companies furnished materials, parts or service in its construction."

The flight was not the heroic lone success of a single daring individual, but the climax of the cooperative effort of an elaborately interlocked technology. The day after Coolidge's speech, Lindbergh said at another ceremony in Washington that the honor should "not go to the pilot alone but to American science and genius which had given years of study to the advancement of aeronautics." "Some things," he said, "should be taken into due consideration in connection with our flight that have not heretofore been given due weight. That is just what made this flight possible. It was not the act of a single pilot. It was the culmination of twenty years of aeronautical research and the assembling together of all that was practicable and best in American aviation." The flight, concluded Lindbergh, "represented American industry."

The worship of the machine which was embodied in the public's response to Lindbergh exalted those very aspects which were denigrated in the celebration of the flight as the work of a heroic individual. Organization and careful method were what lay behind the flight, not individual self-sufficiency and daring romance. One magazine hailed the flight as a "triumph of mechanical engineering." "It is not to be forgotten that this era is the work not so much of brave aviators as of engineers, who have through patient and protracted effort been steadily improving the construction of airplanes." The lesson to be learned from Lindbergh's flight, thought a writer in the *Independent,* "is that the splendid human and material aspects of America need to be organized for the ordinary, matter of fact service of society." The machine meant organization, the careful rationalization of activity of a Ford assembly line, it meant planning, and, if it meant the loss of spontaneous individual action, it meant the material betterment of society. Lindbergh meant not a retreat to the free life of the frontier past but an emergence into the time when "the machine began to take first place in the public mind—the machine and the organization that made its operation possible on a large scale." A poet on this side of the matter wrote, "All day I felt the pull / Of the steel miracle." The machine was not a devilish engine which would enthrall mankind, it was the instrument which would lead to a new paradise. But the direction of history implicit in the machine was toward the future, not the past; the meaning of history was progress, not decline, and America should not lose faith in the future betterment of society. An address by a Harvard professor, picked up by the *Magazine of Business,* made all this explicit. "We commonly take Social Progress for granted," said Edwin F. Gay, "but the doctrine of Social Progress is one of the great revolutionary ideas which have powerfully affected our modern world." There was a danger, however, that the idea "may be in danger of becoming a commonplace or a butt of criticism." The speaker recognized why this might be. America was "worn and disillusioned after the Great War." Logically, contentment should have gone with so optimistic a creed, yet the American people were losing faith. So Lindbergh filled an emotional need even where a need should have been lacking. "He has come like a shining vision to revive the hope of mankind." The high ideals of faith in progress "had almost come to seem like hollow words to us—but now here he is, emblematic of heroes yet to inhabit this world. Our belief in Social Progress is justified symbolically in him."

It is a long flight from New York to Paris; it is a still longer flight from the fact of Lindbergh's achievement to the burden imposed upon it by the imagination of his time. But it is in that further flight that lies the full meaning of Lindbergh. His role was finally a double one. His flight provided an opportunity for the people to project their own emotions into his act and their emotions involved finally two attitudes toward the meaning of their own experience. One view had it that America represented a brief escape from the course of history, an emergence into a new and open world with the self-sufficient individual at its center. The other said that America represented a stage in historical evolution and that its fulfillment lay in the development of society. For one, the meaning of America lay in the past; for the other in the future. For one, the American ideal was an escape from institutions, from the forms of society, and from limitations put upon the free individual; for the other, the American ideal was the elaboration of the complex institutions which made modern society possible, an acceptance of the discipline of the machine, and the achievement of the individual within a context of which he was only a part. The two views were contradictory but both were possible and both were present in the public's reaction to Lindbergh's flight.

The Sunday newspapers announced that Lindbergh had reached Paris and in the very issue whose front pages were covered with Lindbergh's story the magazine section of the *New York Times* featured an article by the British philosopher, Bertrand Russell. The magazine had, of course, been made up too far in advance to take advantage of the news about Lindbergh. Yet, in a prophetic way, Russell's article was about Lindbergh. Russell hailed the rise to power of the United States because he felt that in the "new life that is America's" in the 20th century "the new outlook appropriate to machinery [would] become more completely dominant than in the old world." Russell sensed that some might be unwilling to accept the machine, but "whether we like this new outlook or not," he wrote, "is of little importance." Why one might not was obvious. A society built on the machine, said Russell, meant "the diminution in the value and independence of the individual. Great enterprises tend more and more to be collective, and in an industrialized world the interference of the community with the individual must be more intense." Russell realized that while the cooperative effort involved in machine technology makes man collectively more lordly, it makes the individual more submissive. "I do not see how it is to be avoided," he concluded.

People are not philosophers. They did not see how the conflict between a machine society and the free individual was to be avoided either. But neither were they ready to accept the philosopher's statement of the problem. In Lindbergh, the people celebrated both the self-sufficient individual and the machine. Americans still celebrate both. We cherish the individualism of the American creed at the same time that we worship the machine which increasingly enforces collectivized behavior. Whether we can have both, the freedom of the individual and the power of an organized society, is a question that still haunts our minds. To resolve the conflict that is present in America's celebration of Lindbergh in 1927 is still the task of America.

19

KEATON AND CHAPLIN: THE SILENT FILM'S RESPONSE TO TECHNOLOGY

George Basalla

Any discussion of movies and modern technology is certain to focus upon Charlie Chaplin and the problems he encountered as a factory worker in the film *Modern Times* (1936). American cinematic reaction to technology, however, began prior to Chaplin's portrayal of a beleaguered factory hand and it can be fully understood only if we consider the mechanical nature of film itself, and the motion picture comic tradition that preceded *Modern Times*.

Modern Times, as well as any other film, comic or tragic, silent or sound, live or animated, must be considered in relation to one of the more significant developments of 19th-century culture: the mechanization of the arts. Beginning with the invention of photography in the 1840's, technology posed a threat to traditional artistic practices and forms, while at the same time it raised the possibility of novel techniques and wholly new art forms.

Arguments about the artistic status of photography and the role of the photograph as a source of inspiration for the painter or sculptor had not subsided when the phonograph appeared in the 1870's. Although the phonograph did not immediately raise questions about the creation of a new art form, as did the camera, it could be cited with photography as proof that artistic works, no matter how subtle or sublime, were amenable to *mechanical reproduction*. The advent of the camera and phonograph proved that art could be multiplied by a machine and prepared for wide distribution to the masses.

The culmination of the movement towards the mechanization of the arts was reached with the invention of motion pictures in the final decades of the 19th century. In cinema mechanical civilization found its natural mode of artistic expression. It was an art form that was in essence mechanical, industrial, and commercial—a fact which is reflected in the characterization of early films as "canned" drama or machine-made theater. Among other things, the motion picture is a modern industrial product that results from the cooperative efforts of a group of artists, entrepreneurs, and technicians. To extend the industrial metaphor, it has been said that these makers of movies assemble their product in a manner analogous to that used on an automobile assembly line. Prior to the development of the cinema, no art form had ever been so dependent upon the mechanical and commercial world for its full realization.

Who were the first producers and consumers of the cinematic illusions created by the Hollywood film industry? The early movie moguls were, for the most part, self-made men, immigrant entrepreneurs who founded the film industry and then continued to dominate it throughout most of the 20th century. Initially they provided entertainment for an industrial working-class audience that could afford five or ten cents for a movie ticket but not the dollar or two needed for admittance to a performance at a legitimate theater or opera.

Technology, so intimately associated with the creation of moving pictures and their early audience, also accounted for an important part of the actual stories and images of the film. Arnold Hauser in his *Social History of Art* claims that because cinema was founded in technology, it is at its best in describing "movement, speed, and pace." Silent filmmakers searched their rural and urban landscape for things that moved with speed, and they found them in abundance: steam locomotives, motorcycles, steamships, airplanes, trolley cars, and automobiles. Long before there was developed an explicit aesthetic of the machine, filmmakers made use of the engines of industrial civilization. They elevated machines to the center of cinematic interest and in some notable cases managed to make them the stars of the moving picture show.

Virtually all of the early slapstick comedies, especially those of Mack Sennett, incorporated hectic automobile-locomotive-trolley-etc. chase scenes. Some of the early filmmakers, entranced with the depiction of speed on the screen, offered films that were merely melodramatic, comic, or mechanical embellishments of a prolonged chase. Whether their goal was comedy or tragedy, the men who made silent films concentrated upon moving objects so essential to their art form and thus developed a cinema of speed and exhilaration that has remained as a crucial part of modern cinematic art.

* * *

The foregoing discussion of the early technological background and associations of film art, audience, and content forms a proper introduction to a close examination of two dominant film personalities who dealt with the machine and industrial society in their comedies: Buster Keaton and Charlie Chaplin.

Throughout his life Buster Keaton was an inventor, designer, and builder of machines. Some of the devices were props for his

Buster Keaton sits undaunted as his automobile disintegrates around him.

Source: Film Stills Archives/Museum of Modern Art, New York

films, others were constructed as practical jokes (an outhouse with collapsing walls), and still others were absurd or fanciful contraptions in the style of Rube Goldberg. Keaton's fantastic machines reveal his understanding of the mechanical—the gadgets he built worked—his appreciation of mechanical motion for its own sake, and his ability to use the machine as a source of gentle, and muted, satire of a society that went to great technological lengths in order to accomplish the simple or the unnecessary.

Between 1924 and 1926 Keaton released two films which dealt directly with some major examples of technology. In *The Navigator* (1924) Buster Keaton turned a large passenger liner into a nemesis that threatened the comic figure he created on the screen. A pampered, incompetent, young millionaire and his reluctant lady-love are cast adrift, alone, on the 500-foot steam vessel *Navigator*. Without crew, fuel, or electricity the ocean liner becomes a monster, a "mechanical whale," with Jonah-Keaton and friend trapped inside. Food is available in abundance, but packaged for two hundred, not two; modern cookingware abounds but its capacities are measured in gallons, not cups. The darkened ship at night is as ominous a setting as a haunted house. But resourceful Buster, meeting the challenge of the machine, creates a hospitable environment aboard, especially in the kitchen where a set of his ingenious inventions enabled the couple to get the food they needed.

Keaton's biographer, Rudi Blesh, has said that *The Navigator* is not the story of romance impeded by the machine, but rather "the tale of a precarious wooing that hinges on [Keaton's] duel with a mechanical enemy." There is no doubt of the machine's hostility in *The Navigator,* but neither is there doubt that Buster will deal successfully with technology's challenge to the individual. The little guy always manages to outwit the big machine.

The General (1926) suggested a different emphasis in Keaton's continuing search for a solution to the personal problems engendered by the machines people meet in their lives. Set in the Civil War period, the film was based on the true story of a daring, but unsuccessful attempt by Northern raiders in 1862 to infiltrate Southern territory and disrupt rail service between Georgia and Tennessee. The Northern plan included the capture of a locomotive, the "General," that the raiders would use as they moved northward burning bridges and cutting communication lines.

Since the "General" was a Southern locomotive, Keaton saw the necessity of telling the story from the Confederate viewpoint in order to elicit sympathy for engine and engineer. Keaton was best able to exploit the full cinematic and comic possibilities of the story not only because he expressed Confederate sympathies but also because he made his hero a civilian closely associated with the film's mechanical co-star. This move called for a more harmonious relationship between man and machine. The "General" is no *Navigator* beyond the control of those aboard; it is a love object, normally obedient to man's will, that is lost and recovered.

The "General" and Buster's girl friend, Annebelle Lee, are both loved by him; yet their individual and collective actions can be frustrating to our hero. Neither is malicious, but neither is entirely cooperative. The girl is a bit scatterbrained, and the engine, often when under the girl's control, works contrary to Keaton's best interests and intentions. And when girl and engine cooperate, other mechanical malfunctions arise. Buster is able to handle these malfunctions, but only in the nick of time.

It is not the machine's hostility towards people at issue here, it is that mechanical and human needs and desires do not mesh smoothly, they are slightly out of phase and require a Keaton, with wit and ingenuity, rather than superior technical knowledge, to set them right. All of this might be summarized in a moral derived from Keaton's films: if people are to prevail over technology they must be wary, even of the machines they love.

* * *

Buster Keaton, who had no artistic successes with sound films, had shown a preference in his best works for historical settings —the Civil War—and the machines of the 19th century—steam engines. His primary concern was not the social, economic, and technological problems of 20th-century America. Charlie Chaplin, who gained fame in both silent and sound movies, made what is probably the most famous film commenting upon 20th-century technology: *Modern Times* (1936).

Chaplin was motivated to make a film on the role of the machine in modern society by a young reporter who, while interviewing him, happened to describe working conditions in industrial Detroit. He was told that healthy, young farm boys were lured to the city to work on the assembly line producing automobiles. Within four or five years these same young men were "nervous wrecks," their health destroyed by the pace of work set in the factory. The reporter's observations reinforced Chaplin's own "impulse to say something about the way life is being standardized and chan-nelized, and men turned into machines...."

Chaplin did not share Keaton's intimate knowledge of the machine, but he did aspire to, and move in, intellectual and social circles where technology might be the subject of abstract discussion. Silent films had made Chaplin an international figure with access to the greatest minds of the day. The combination of his own poverty-stricken childhood and the influence of the more socially-concerned intellectuals he met, men like Max Eastman and H.G. Wells, turned multimillionaire Chaplin into a parlor socialist who enjoyed discussing social, economic, and political matters with radical overtones.

Apart from any expression it might find in *Modern Times,* what were Chaplin's attitude towards labor and the machine? The answer must be a confusing and unsatisfactory one, for it is derived from Chaplin's occasional excursions into the realms of economics and political philosophy. The standard Chaplin solution for all economic problems was: "Reduce the hours of labor, print more money, and control prices." During the American Depression he lectured Albert Einstein along these lines and was pleased when the physicist replied, "You are not a comedian. You're an economist." The Einstein conversation also found Chaplin claiming that technological unemployment—new machines displacing men from their jobs—was at the heart of America's economic problems in the 1930's. Yet, when Chaplin had an interview with Mahatma Gandhi, he made clear his sympathy with the drive for India's independence but expressed confusion at the Indian leader's "abhorrence of machinery." Gandhi patiently explained to the comedian that India could only gain political independence if she first reduced her technological dependence upon Great Britain. In all of this there is nothing that is profound or novel. Chaplin was merely passing on some of the liberal clichés of his day and doing it in a naïve and superficial manner.

Modern Times, as described in its foreword by Chaplin, "...is the story of industry, of individual enterprise—humanity crusading in the pursuit of happiness." The opening scene discloses a mass of sheep, jammed into pens, presumably on their way to the abattoir. Abruptly this scene cuts to a similar one of a crowd of factory workers jostling one another on their way to work in an industrial plant. The sharp note of social criticism implied by this sheep/worker montage is not sustained throughout the film. It is replaced by a more gentle tone of satire, by an obvious desire to please the audience and not to lecture to it.

Caught up in the giant cogs of a machine, Charlie Chaplin personifies the dehumanizing effects of mass production in Modern Times *(1936).*

Source: Film Stills Archives/Museum of Modern Art, New York

Scenes of factory interiors account for less than one-third of the footage of *Modern Times,* but they contain some of the most pointed social commentary as well as the funniest comic situations. Everyone who has seen the film recalls Charlie vainly trying to adapt to the pace of the fast-moving conveyor belt, literally going mad in the process and running about the plant tightening real and imaginary nuts with his wrench and spraying oil on factory personnel.

Another popular scene featured a newly invented automatic feeding machine brought to the assembly line so the worker need not interrupt his labors while he ate. The feeding machine malfunctioned, hurling soup, meat, corn-on-the-cob, and a napkin at Charlie, who was strapped into the contraption and could not escape. In all of cinema there is no finer scene illustrating man's utter help-lessness before the machine that supposedly exists only to serve his most basic needs.

The bulk of *Modern Times* was set beyond factory walls in the institutions and business establishments of an industrial town. In this way we learn not only about the assembly line but the social and home life of the workingman. The comedy continues in the new setting but the scene is a grim, depressing one. Charlie leaves the factory to recuperate in a hospital; from there he goes to prison. A strike by the factory workers brings with it violence, death, and unemployment—all of which impinges upon Charlie's life. Mechanical transportation for Charlie takes the form of ambu-lances and police prison vans, not the automobile so dear to the American worker's dream. The one idyllic interlude in the film finds Charlie with a young girl he befriends and protects. They play at housekeeping in an abandoned waterfront shack, mocking middle-class respectability and pretensions. The intervention of the juvenile authorities seeking the girl ends their happiness, and at the film's conclusion Charlie and friend tramp down a lonely road into the sunset. They have been rebuffed and harassed by every facet of industrial society; their only hope lies in escaping from it. They leave behind them maddening work on the assembly line, unemployment, strikes, poverty, and social injustice. Chaplin had no ready solutions to these problems.

Film critics of the day tended to respond to *Modern Times* according to their preconceived notions of Chaplin and socialism. A writer for the Communist *New Masses* praised it as the first American film "daring to challenge the superiority of an industrial civilization based upon the creed of men who sit at flat topped desks and push buttons demanding more speed from tortured employees." Others, more conservative, were relieved to find that it was "neither fish, flesh, nor good red propaganda," that its social message was innocuous and its comedy high, if at times uneven. Over the years critical opinion has tended to stress the derivative nature of *Modern Times*—Chaplin lifted many comic bits from his earlier films—and its episodic character.

Clearly *Modern Times* has its faults, both of an artistic and ideological sort, but it remains the single best film ever made for a mass audience treating technology within a social context. It did not offer a radical social message for those wishing to overthrow capitalism; it did, however, accurately reflect the sentiments of many who felt they were helpless victims of an over-mechanized world. And finally it did conceal a truly disturbing message beneath its wonderful comic scenes: machines cannot be mastered by human beings. The neat tricks that Buster Keaton used to keep his machines in line simply have no place in Chaplin's *Modern Times*. The only solution is to turn one's back on the whole affair and walk down the road with a friend. But over the next hill we know that Charlie is likely to find waiting for him another factory, another industrial town, another set of problems generated by the social impact of technology.

Charlie on the screen might decide to turn away from the machine in desperation and seek happiness elsewhere. However, Chaplin as a film director had to determine exactly how he would handle the technology of sound in *Modern Times*. The talkie was almost a decade old, and it had eclipsed the silent film. In a *New York Times* article of 1931, Chaplin argued that talking pictures could never serve as adequate substitutes for silent films based upon the universal language of pantomime. He ignored lip-synchronized sound when he made *City Lights* (1931) and went on to issue *Modern Times* five years later as the last full-length silent feature film made in Hollywood.

There is a non-dialogue, musical sound track for *Modern Times*, but the film was shot for the most part at silent speed. At three specific points in the film recognizable human speech was permitted to intrude into the otherwise pantomimed action: first, when the factory's top executive used *closed-circuit television* to talk to his employees in the plant; second, when the inventor of the feeding machine played a *phonograph record* extolling the virtues of his device; third, when a *radio* broadcasts a news bulletin in the prison scene. By limiting the sound of the human voice to three technological devices capable of transmitting and reproducing it, he underlined the technological character of the talkies that had supplanted the silent movies. The talking motion picture, he appeared to be saying, is but another technical gadget; the true value of a film's drama lies in the pantomimic art of the actors and actresses.

Even Chaplin, the master of pantomime, was forced to bow to what most people then saw as inevitable technological progress in the cinema. Near the conclusion of *Modern Times*, Charlie sang a song that had deliberately contrived gibberish lyrics and a great deal of pantomime. Nevertheless, film goers heard Charlie Chaplin's voice for the first time over the loudspeakers in a movie theater. After that there could be no going back, no walking off into the sunset. If Chaplin were to produce another film, it must be a full-fledged talking picture. *The Great Dictator,* appearing four years later, was such a picture.

* * *

Technology was not a subject matter haphazardly introduced into the silent film by directors who were willing to include in their screen productions anything they happened upon. The machine, in mechanizing the arts, gave birth to the cinema, which was especially suited to the needs and desires of industrial society; the machine deeply influenced film content and acting style; the machine brought sound to the silent movie and altered the future direction of cinematic development. Similarly, it was not by chance that the two greatest silent film comedians, Keaton and Chaplin, used cinema to explore the nature of technology, man, and society. They were using a mechanical art form in an attempt to understand the changing Machine Age in which they lived. What is ironic in all of this is that a medium of artistic communication closely allied to the machine did not produce film masterpieces praising technology but ones that were suspicious and critical of the machine and the civilization it spawned.

20

MORRIS L. COOKE AND ENERGY
FOR AMERICA

Jean Christie

On September 4, 1882 on Pearl Street in lower Manhattan, the
first central electric power station sent current out over its wires.
Its customers were owners of some 400 of the carbon filament light
bulbs recently invented by Thomas A. Edison, but Edison, founder and director of the company, insisted that electricity would
soon find a market not only for lighting but for many other uses as
well. Not long afterward, advances in the transmission of alternating current enlarged the territory of central stations, and within 30
or 40 years electric light superseded gas, electric trolley cars
transported passengers within and between the growing cities, and
electric motors replaced moving belts in the factories of rapidly
industrializing America. Although coal-burning plants generated
most of the current, electric companies, attracted by the potential
of hydro-electric power, also sought to obtain promising sites on
the nation's rivers. Engineers constructed larger plants and longer
transmission lines, and bankers, entrepreneurs, and speculators
organized complex structures to carry on an expanding business.
In a capitalistic economy, electrical service was, as a matter of
course, undertaken for profit, and although some cities set up their
own municipal systems, privately owned companies predomi-

nated. These corporations held exclusive franchises to supply what was coming to be regarded not as a luxury but as an essential of life.

In the late 19th century, indeed, concentration had replaced competition throughout so much of the economy that, amid exultation at the marvels of technology and the outpouring of material goods, there arose social and political movements demanding popular control over the great combinations or "trusts." The quasi-radical Populist Party went down to defeat in the 1890's, but in the early 20th century millions of Americans became uneasily aware that headlong exploitation of material resources and of human labor had entailed waste, confusion, and social injustice. Although the Socialist Party gained adherents, moderate reformers known as "Progressives" dominated the political scene. Never a coherent entity, the Progressive "movement" consisted of diverse individuals and groups who focussed their attention on a variety of social and economic evils for which they proposed a variety of remedies. They shared, however, a belief that "the people" should regain control of affairs and that government —local, state or federal—should initiate action for "the public interest."

A number of Progressives concentrated on the question of how to realize the full potential of electrical technology for the benefit of everyone. Among these reformers, Morris L. Cooke of Philadelphia played a prominent part. Cooke, a mechanical engineer, was also a disciple of Frederick W. Taylor, the founder of Scientific Management, which applied systematic formulas to examine and organize the production, sales, and administrative functions of a business. Cooke viewed this system as the key to an unimaginable abundance that would enable human beings to create a more harmonious and more spiritual civilization. Since he possessed both wealth and a sense of public responsibility, he devoted much of his time, both as a private citizen and often as an appointed official, to various attempts to apply the principles of planning and rationality to "the public business." Engineers, he believed, had an inspiring function in society: to liberate mankind though applying science to human needs. Like his acquaintance, economist and social critic Thorstein Veblen, he sharply distinguished his profession from profit-oriented enterprise and protested "against the assumption that business—big or little—is

Engineering.''

To the group of "power progressives" technological advance had resulted from cooperative social effort to master the material world, but a minority had seized the fruits for themselves. Progressives denounced the attempts of corporations to gain control over water power sites, accused them of gouging consumers through exorbitant rates, and charged that in pursuit of the largest possible profit they skimmed the cream of the market and neglected to serve the rural population. Enthusiasts for technology, the progressives maintained that all should be able to use electricity freely and looked forward, with Cooke, to the era predicted by electrical engineer Charles Steinmetz, when electricity "is going to be so cheap that it is not going to pay to meter it." They differed among themselves on the most effective method to hasten that day. All agreed that government must take action, but while some demanded public ownership of all electrical systems, others believed that regulation by public commissions would be sufficient, and yet others proposed public ownership of certain facilities, particularly in hydropower, that could serve as "yardsticks" to measure the performance of the private industry. Essentially, Cooke held the last view. Though frequently critical of the regulatory commissions, he refused to commit himself to the cause of full public ownership but enthusiastically supported several major government power projects.

Up until the First World War, which the United States entered in April 1917, these reformers had made progress in raising the power issue for discussion, often in connection with the movement for conservation of resources, and had tasted victory in certain specific cases—Cooke, for instance, as Director of Public Works in Philadelphia, had forced the local utility company to reduce its rates somewhat—but they found after the war that the wave of reform had subsided and that conservative business interests again held firm control of government and set the tone of political discussion. As Washington administrations looked on benignly at the increasing concentration of business, the flourishing electrical industry, especially, was rapidly coming into the hands of promoters who gathered local units into pyramids of holding companies that often milked the operating companies. "The power combine," lamented the power progressive Judson King, "now is more powerful politically than the railroads."

Technology had advanced so that by 1920 engineers could transmit high-voltage current over distances of 200 miles or more, and many claimed that huge "superpower" generating stations and long-distance transmission lines could link extensive territories into continuous networks. Yet inexpensive power remained beyond the reach of millions of people. While city dwellers grumbled at the cost of current to light their homes and run their toasters, vacuum cleaners, and washing machines, ordinary country people had no electricity at all. Rural families still lit their houses and barns by kerosene lamps and carried on domestic tasks by exerting human muscle.

Though the political situation was discouraging, power progressives stuck to their purpose and doggedly fought on. J. D. Ross, head of Seattle's municipal system, warded off encroachments from private companies; Judson King, who had adopted the cause as one in which he could "help construct one or two new steps in the stair-way of justice," lobbied in Washington for the National Popular Government League. United States Senator George W. Norris of Nebraska led a campaign to have the federal government develop hydropower at an unfinished wartime plant at Muscle Shoals on the Tennessee River. Other progressive (or "liberal") politicians, such as Franklin D. Roosevelt of New York and Gifford Pinchot, a combative conservationist and former Chief Forester of the United States, also worked to achieve state or federal control over the utility companies and to make cheap power widely available.

Financially, socially, and professionally secure, Morris L. Cooke in his 50's enjoyed good health and an overflowing supply of energy. In later life he told a young friend: "I advise you always to have twenty causes you are working for: nineteen will fail, but one will succeed." For years Cooke had been engaged in a running battle against utility company influence in the professional engineering societies, during which he had been officially censured by the council of his own American Society of Mechanical Engineers. In the early 1920's he led a temporarily successful, but short-lived, revolt of the liberal engineers. An apparent refusal of the societies to permit organized discussion of electric rates led him to believe that "the engineering profession in being silent in the face of the obvious facts is almost making itself a party to a conspiracy, which bears especially hard on the women of the land." He made his office a center of information and encouragement for the circle of activists in the cause, and personally initiated the financed studies of the cost of distribution of electricity to domestic consumers, that is, from the local station to the place of actual use. This was a touchy point with the companies, since they defended their domestic rates as necessitated by unavoidably high costs of distribution.

In 1922, when Gifford Pinchot won election as Governor of Pennsylvania, he called on Cooke to advise him on utility questions. Cooke proposed, and himself directed, a Giant Power Survey Board, authorized not only to study the water and fuel resources of the state but also to recommend "such policy with respect to the generation and distribution of electric energy as will . . . best secure for the industries, railroads, farms, and homes of this Commonwealth an abundant and cheap supply of electric current. . . ." In its report, *Giant Power* (1925), the Board presented a plan to control the corporations, to utilize Pennsylvania's coal resources, and to make full use of the latest technology. The state, it proposed, would direct a reorganization of the industry and would license "mine-mouth plants"—power-generating stations located next to the mines—in the bituminous fields of western Pennsylvania, from which "coal by wire" would radiate on long lines to industries and to rural households throughout the state. The project, Pinchot told the legislature, could lift "most of the drudgery of human life . . . from the shoulders of men and women who toil . . . ." A gigantic monopoly, he warned, might soon control "the greatest material blessing in human history." The people must act quickly, for "either we must control electric power, or its masters and owners will control us." But his plea fell on unheeding ears, and the plan never reached the floor for debate.

In New York state Franklin D. Roosevelt became governor in 1929 and in his turn called on Cooke for counsel. His predecessor, Al Smith, had preserved state water power sites in public hands and now Roosevelt called for "the utilization of this stupendous heritage." He appointed Cooke to the new Power Authority of the State of New York to plan for the disposition of vast supplies of current which they (mistakenly) believed would very soon be provided by the St. Lawrence-Great Lakes project. At Cooke's instigation the Authority studied distribution costs. Confronted with inadequate data—for the industry gave out only sketchy information—Cooke reminded the staff that their purpose was not to reach unassailable conclusions but to prove a case for action. "We can once and for all discredit the present system of domestic electric rates," he asserted. "A large part of our job is to inspire the people of the State generally to a realization of what cheap and abundant electricity will mean." Early in 1933 the Authority sponsored a conference on domestic rates. Noting that smaller bills would encourage people to increase consumption and that such heavier purchases would lower costs per unit of electricity, Authority experts reported that at current average usage the rates might well be lowered from the prevalent 7 cents to 3 or 4 cents per kilowatt hour. Tentative though they were, the figures were solid enough to impress some spokesmen of the industry.

Considering the circumstances, the power progressives made remarkable headway during the 1920's. So blatant were the holding companies, and so desirable was cheap energy, that the reformers in this field could count on widespread, if unorganized, support. Twice they gained Congressional approval (frustrated by presidential vetoes) of the Muscle Shoals proposal. They initiated state and federal investigations, and they developed the power question into a prominent political issue.

In the Great Depression that began in 1929, holding-company empires collapsed, big business in general was discredited, and, led by Franklin D. Roosevelt, a Democratic administration initiated the period of liberal reform known as the New Deal. It was a heady time for the power progressives. Conservatives themselves clamored for government action, while intellectuals and administrators advocated economic and social planning. The new President's record and some of his appointments, like that of Chicago progressive Harold L. Ickes as Secretary of the Interior, suggested that the New Deal might open a new electrical era.

At the outset Roosevelt adopted and enlarged the scope of the Norris proposal for Muscle Shoals; he wanted to plan for the whole Tennessee Valley, "tying in industry and agriculture and forestry and flood prevention, tying them all into a unified whole over a distance of a thousand miles so that we can afford better opportunities and better places for living for millions of yet unborn in the days to come." In May 1933 Congress assigned to the Tennessee Valley Authority a mandate to develop the river and a broad responsibility for the welfare of the inhabitants of the entire Ten-

The Guntersville Dam on the Tennessee River, built by the TVA in 1935–39, creates a reservoir for flood regulation and provides some 97,000 kilowatts of installed generating capacity.

nessee Valley. With Hoover (or Boulder) Dam already under way, the government constructed other major hydro projects, such as Bonneville and Grand Coulee, that today furnish current throughout a vast area of the Pacific Northwest. A revitalized Federal Power Commission surveyed power resources and costs and reached conclusions similar to those of the New York Authority. This kind of pressure, often repeated on the local level, brought results in a general lowering of electric rates. Congress, moreover, tackled the organization of the industry when it passed the Public Utility Holding Company Act of 1935, which was intended to bring the corporations under more effective regulation and to eliminate those holding companies that served no useful service function.

Cooke jubilantly participated in most of these activities, yet saw them as only the beginning of what should be done. More technically minded than many of his fellows, who rejected any contacts with the industry, he advocated "power pooling" to create regional networks of transmission lines to carry energy from sources both public and private. He was convinced, furthermore, that the national government ought to set forth the principles of a long-range policy to guide the development of energy for the future. In 1934, when the President selected him as one member of an advisory National Power Policy Committee, he looked forward to drawing up such a statement. He was disappointed; although the group was largely responsible for the Public Utility Holding Company Act, it proved to be too deeply divided even to make headway on the broader task.

In the fall of 1933 Cooke was appointed to chair a "Mississippi Valley Committee." Though its original function was to advise on a mass of applications for federal aid to public works projects, he saw a considerable chance that it could "become the nucleus of an ultimate scheme of national planning . . . ." The Committee drew up a proposal for the whole interior watershed of the United States. The members hoped to coordinate flood control, soil conservation (for which Cooke had become a passionate propagandist), hydropower and rural electrification, even land use and the preservation of the wilderness. Later, Cooke chaired a committee on the drought-stricken Great Plains and, after the Second World War, a committee on water resources policy. All plans, he insisted, must take into account a complex of interrelated forces, and by the late 1930's he had "discovered" the scientific field of ecology and was pleading for "total conservation."

But the administration's interest in planning, never profound, had largely evaporated after the first year or two of the New Deal. No subsequent project was charged with the responsibilities of the TVA, and that agency itself came to be chiefly a supplier of electricity. Cooke's proposals for regional and coordinated development, since they implied quite centralized national control, aroused anti-bureaucratic fears and offended a host of particular interests—rival government agencies, large corporations, and local landowners and entrepreneurs. A number of features were separately adopted, but the vision of watershed planning never gained acceptance.

It was to rural America that Cooke made his greatest concrete contribution. In the 1920's he had reproached the electrical industry for its failure to serve the countryside, and by the 1930's he could point out that in a decade the proportion of farms supplied with central station electricity had risen only from 1.6 percent in 1920 to 10.4 percent in 1930.* From the earliest days of the New

---

*Another three percent of farms had their own generating plants.

Deal he prodded officials and fired specific proposals at Secretary Ickes and the President. Others were pushing in the same direction: Senator Norris, of course; farm organizations; TVA and pioneer cooperative groups in the territory that it supplied with current. Finally, in the spring of 1935 the President, who had been provided with a large fund for unemployment relief, allocated a hundred million dollars to set up a Rural Electrification Administration and appointed Cooke as its head.

The agency began amidst confusion: it was to lend money to construct distribution lines—but to whom? Private companies were willing to take the money, but would accept no government control over the rate structure. After a difficult year cooperatives emerged as a practical solution, and under the forceful leadership of John Carmody, who in 1937 succeeded Cooke as Administrator, they multiplied rapidly. With REA assistance, rural neighbors joined together, borrowed funds at moderate interest from REA, contracted (usually) for power from the nearest supplier, and built the lines to light their houses, pump the water, milk the cows. REA devised a number of improvements to cut down costs, and it insisted on ''area coverage'' in order to form a compact body of customers in the territory of each cooperative. Low rates would encourage higher consumption of current. Ridiculing "the electrified farmer in the New Deal dell," utility companies tried to stop

With the establishment of the Rural Electrification Administration in 1935, power reached into farmlands like these, where linesmen are stringing electric lines.

REA through court actions and in some territories attempted to forestall the cooperatives by rapidly stringing lines to the more prosperous residents—a tactic that resulted in "the battle of the shovel." As REA nevertheless made headway, the companies began to extend their rural service in earnest. By the mid-1950's, 96 percent of rural dwellings had access to electrical energy. Those who have lived without that convenience can best understand the difference it made.

The power progressives enjoyed their greatest influence during the five or six years of the New Deal. By 1939, and in the ensuing years of war, the demand for energy to run war industries overshadowed innovation or consideration of long-range policy. While REA marked time, TVA expanded enormously as supplier of power. After the war the liberals proposed a Missouri Valley Authority, but without success; Cooke chaired the President's Water Resources Policy Committee (his last public office), but President Harry Truman refused to take action on its three-volume report and its recommendations for a system of planning for "ten rivers in America's future." The reform period had ended, and conservative forces again ruled almost unchallenged.

Technologically, the nation enjoyed a tremendous expansion of energy facilities. Although creations of the New Deal such as the great hydro projects and REA had become too popular to dismantle, the power companies had recovered their position, and in the 1950's a Republican administration hailed the "partnership" of government and industry. Networks of high voltage transmission lines covered major areas of the country, and in the 1960's minemouth plants, heralded by Cooke in the 1920's, converted the coal of southwestern deserts into power for cities of the West Coast. With nuclear energy on the horizon, it seemed clear to Cooke's friend Leland Olds, one-time member of the Federal Power Commission, that "low-cost electricity is an inexhaustible resource," and he urged public systems to structure their rates so as to reward consumers who used current most heavily in all-electric homes. Millions of Americans indeed took for granted the flow of electricity, whether to manufacture goods, cook their meals, or light, heat and cool their houses.

In the 1960's movements for social change arose again in America and reopened a wide range of issues, among them questions of energy. In the 1970's a growing number of critics are discussing the situation in terms that in some ways echo, in other ways contradict, the progressives of an earlier day. In language reminiscent of the 1920's and 1930's, consumers inveigh against monopolistic corporations and against high utility rates; this time, however, they are more likely to find cheaper public electric service within their reach, obtainable, perhaps, from the TVA or the New York State Power Authority. But many searching critics sound a far different note from the power progressives, for they reject the goal of unending expansion. Resources of energy are not inexhaustible, they declare. Pointing out that the strip-mining of coal devastates the land and that mine-mouth plants pollute the desert air, they insist that nuclear plants endanger the lives of hundreds of thousands of people. Therefore, they call on Americans to reverse direction and to set themselves to consume not more, but less, electrical energy. They demand, indeed, a reconsideration of the purposes and direction of technology.

Though in their generation the power progressives, eager to disseminate the genuine benefits of "giant power," failed to foresee all the consequences of success, they also maintained that a society which has created the marvels of technology can direct those achievements to enhance the well-being of both the people and their environment. With all his zeal for electrical development, Morris L. Cooke insisted also on the integration of the resources of the soil, of the rivers, and of human knowledge into a policy for the nation.

21

ENRICO FERMI AND THE DEVELOPMENT OF NUCLEAR ENERGY

Lawrence Badash

Speaking in the flush of early excitement over the nuclear reactions induced in particle accelerators, the great British scientist Ernest Rutherford cautioned that commercial utilization was virtually impossible because the energy released in these reactions was far less than that needed to run the machines producing them. "Anyone who expects a source of power from the transformation of these atoms is talking moonshine," he warned in a 1933 *New York Herald Tribune* interview.

Exploring What Holds Matter Together

The origin of the concept of atomic energy is shrouded in antiquity. Ancient Greek ideas of elements included the understanding that other materials were made of combinations of the elements. This raised the question of what held the elements together in these combinations. Aristotle's four-element theory may seem simplistic, but one interpretation of it has the earth, water, and air representing the solid, liquid, and gaseous states of matter, while fire was the symbol for energy. Thus, taking some historical liberties, we might argue that Aristotle's energy, not points or hooks, was the glue that held together the atoms of Democritus.

On much more solid ground, and much closer to the present time, we may look to the early 19th-century experiments in which substances were electrochemically dissociated into their constituents. This logically suggested that atoms were joined by electrical forces. Later in the century came the great laws of thermodynamics and conservation of energy and the kinetic theory of gases. These advances gave understanding to the different manifestations of energy, quantified them, and explained that the energy of motion of gas molecules depended on their temperature.

The discovery of radioactivity by Henri Becquerel in 1896 raised to a critical level the question of just what was occurring within the atom or molecule. For Becquerel found that uranium emitted a powerful radiation and appeared to do so continuously. Did this energy come from the motion of particles, the breaking of bonds, or from some other source?

In Paris in 1903 Pierre Curie and Albert Laborde found that a sample of radium maintained itself at a temperature higher than its surroundings, a most curious circumstance. Curie's statement that "he would not care to trust himself in a room with a kilogram of pure radium, as it would doubtless destroy his eyesight and burn all the skin off his body, and probably kill him" was quoted widely in the press and added to the public fascination with radium. Across the Atlantic, in Montreal, Ernest Rutherford and Frederick Soddy in 1902-03 succeeded in explaining the phenomenon of radioactivity. Atoms of uranium, thorium, radium, and other related elements were, they said, being spontaneously transmuted into other atoms. The process involved the expulsion of some types of particles: the beta particle, which was already known to be an electron, and the alpha particle, which would be identified in a few years as a charged helium atom. These chunks of matter were ejected with great velocity, and the relatively massive alpha therefore carried considerable kinetic energy. It was the loss of this energy, in collisions with neighboring atoms and molecules, that raised the substance's temperature, as found by Curie and Laborde.

If the mechanics of the energy expulsion were somewhat understood, the reason for the radioactive decay was not. This touch of mystery excited editorial writers and promised future insights into the nature of atoms. But the prodigious production of heat, a more tangible concept, became the hook on which to hang the worldwide popularity of radioactivity, and especially its most energetic substance, radium. Marie Curie became "Our Lady of Radium," poems were written about her, and someone even asked permission to name a race horse after her. Radium lamps to light bicycles were felt to be just around the corner, while pea-sized fuel sources to send ships across the ocean were deemed only slightly further in the future. Radioactive materials were to provide a cornucopia of energy. Yet this seemingly unlimited source posed serious conflict with the law of conservation of energy. Where did the energy come from? Sir William Crookes suggested that radio-elements selectively filter out the faster moving air molecules and extract their excess energy. The Curies preferred to think of an undetected radiation that pervaded space, whose presence was known only by the alpha, beta, and gamma radiation it stimulated heavy elements to emit. The German team of Julius Elster and Hans Geitel, as well as Rutherford and Soddy, were closer to the mark in believing that the atom itself possessed the energy, which was released when the atom experienced an internal rearrangement.

But could the energy be released on demand and not just at the rate characteristic of each radioelement, its half life? One line of research attempted to alter this decay rate, to speed it up or slow it down. Extremes of temperature and pressure, and different chemical combinations of the radioelements were tried to no avail. Another approach was a series of attempts to split the atom itself, no doubt anticipating some sort of energy change. Rutherford's former teacher in Cambridge's Cavendish Laboratory, J.J. Thompson, was reportedly "so anxious to bust atoms artificially that ... he would have tried it with a cold-chisel before long." (Bumstead to Rutherford, 30 September 1905) But no one ever found large pieces of atoms broken off as a result of X-ray bombardment, for example. Others subjected atoms to intense electric and magnetic fields, detecting some effects which we now recognize as due to the atom's external electrons, but not irretrievably disrupting the atom itself. Yet the work done in Rutherford's Manchester laboratory by Hans Geiger and Ernest Marsden about 1910 showed that the alpha particles from naturally decaying radioelements were useful projectiles for learning about the targets they struck.

This led to Rutherford's nuclear model of the atom in 1911 and eventually to his disruption of the nitrogen nucleus by alpha bombardment in 1919, producing oxygen and hydrogen. In turn, this led to the conception and development of "atom smashers." In the 1920's Rutherford, now in Cambridge, with James Chadwick successfully transmuted numerous elements in the lower region of the periodic table, using alphas from radioactive sources. But they recognized the need for a more copious beam of projectiles and for greater energy to cause even more transmutations. John Cockcroft and E.T.S. Walton, also students of Rutherford, developed a high-voltage machine with which in 1932 they were able to bombard a lithium target with protons, producing two helium nuclei. Rutherford's work of 1919 deliberately produced nuclear transformations using projectiles from naturally decaying radioelements; the work of Cockcroft and Walton did the same using artificially accelerated projectiles. In this reaction they also verified for the first time Einstein's famous $E = mc^2$ relationship. Another outstanding event in that remarkable year of 1932 was James Chadwick's discovery of the neutron. The neutron, of about the same mass as the proton, was uncharged. This meant that it was not affected by the intense electrical fields surrounding the atom and could strike target nuclei more easily than other projectiles.

This then was the situation at the time Rutherford cast cold water on the idea of extracting usable energy from the atom—or, more precisely, the nucleus—by calling it "moonshine." On the one hand, there were particle accelerators, the most famous of which was Ernest Lawrence's cyclotron developed at the University of California at Berkeley in the early 1930's. It was the prototype of a series of ever-larger machines in which the projectiles were electromagnetically accelerated in a circular path. Cyclotrons—and linear accelerators also—were able to throw protons, electrons, and heavier particles at their targets, but, although some of the reactions released sizeable amounts of energy, the energy input was always greater than the output. The other line of development, inducing reactions with neutrons (which cannot be accelerated since they are unaffected by electromagnetic fields), similarly gave little hope for a net energy output, again because the efficiency of neutron production and neutron-induced reactions was too low. Hence Rutherford's statement, based on existing knowledge at the time, was perfectly accurate. Albert Einstein and Robert Millikan made the same responsible prediction. It took the discovery of a new phenomenon of nature—nuclear fission—to alter the outlook.

Fermi and the Control of Nuclear Fission Reactions

Fission was discovered at the end of 1938 by Rutherford's former student, the radiochemist Otto Hahn, and his colleague Fritz Strassman in Germany. It was explained as the rupture of a uranium atom into two roughly equal pieces when struck by a neutron. The scientific community was fascinated with the news for a number of reasons. First, it was so unexpected, physicists and chemists having previously failed for decades to find atoms breaking off pieces larger than alpha particles. Secondly, the amount of energy released in the fission of a uranium nucleus was enormous—about 200 million electron volts. When this is compared with the one or two electron volts per atom released in the most violent chemical reactions, the significance of nuclear reactions is understandable. And, thirdly, there was the strong likelihood that in the fission process not only were two fragments produced, but extra neutrons might be released as well. If one of these extra neutrons struck another uranium nucleus, causing it to fission, and it in turn released neutrons, one of which hit yet another uranium nucleus, and so on, a sustained chained reaction was possible. If the chain reaction were controlled, we would have a continuous source of energy. If more than one neutron were released in fission, and most of them successfully split uranium nuclei, we would have a geometric increase in the number of fissions. In fact, the process occurs so rapidly—the energy is released so quickly—that we have a bomb.

These facts were recognized by physicists the world over. On no one did they make more of an impression than Enrico Fermi. At an early age Fermi was recognized as Italy's outstanding theoretical physicist, and his colleagues looked to him to restore Italy to its place of honor in the world of science, lost since the days of Galileo, Galvani, and Volta. Brilliant work on the theory of beta particle decay and in the creation of "Fermi statistics" was followed by an uncommon, but thoroughly successful, switch to experimental physics. During the mid-1930's Fermi examined neutron-induced reactions extensively and discovered that, unlike the case with charged projectiles, *slow* neutrons are more effective in causing reactions than those moving with high velocity.

Fermi had used the occasion of the Nobel Prize awarded to him in 1938 to leave Mussolini's Italy in order to settle in the United States. First at Columbia University and then at the University of Chicago, Fermi, Leo Szilard, and others showed that certain materials, such as extremely pure carbon and heavy water, were suitable for slowing the neutrons but not absorbing them. Since heavy water was virtually unobtainable, graphite bricks were used to build larger and larger piles, or reactors. Distributed among the

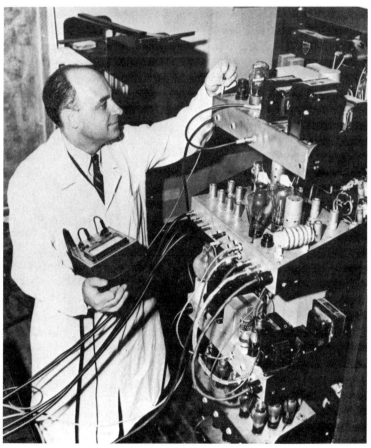

Enrico Fermi works in the physics laboratory at the University of Chicago, where he and his associates achieved the first controlled release of nuclear energy on December 2, 1942.

The first self-sustaining nuclear reactor, partially completed in this photograph, was a crude assembly of graphite bricks and uranium built under the stands of a University of Chicago athletic field.

bricks were lumps of uranium metal and uranium oxide; the fission occurred within the heavy element and the released neutrons lost much of their energy while passing through the carbon moderator before striking other uranium nuclei. On December 2, 1942, in Chicago, Fermi and his associates achieved the first successful controlled release of nuclear energy. Cadmium rods, which absorbed many neutrons, were pulled out of the pile to allow the neutron flux to rise; they were then adjusted to control the neutron level and reinserted to slow or stop the reaction.

But what had this work of a basic and applied scientific nature to do with world events? By this time, of course, the United States had joined the Allies in World War II, and like them was quite fearful that Hitler was making an atomic bomb. This danger has stimulated scientists to present the possible development of nuclear weapons to the American and British governments, and the Manhattan Project was the most significant result. Fermi's success showed three things: a chain reaction was possible, and this meant that a bomb might be made (which is not to say there were no hurdles to overcome here); further, since the part of uranium that fissioned was an isotope (U-235) which existed in the natural metal to the extent of only 0.7 percent and could not be separated from it by chemical means, an entirely new element, not found on earth, could likely be produced in a reactor and this new element would be fissionable. Moreover, the new element, plutonium, would be made by bombarding the otherwise wasted isotope of uranium (U-238) with neutrons and would be separable by chemical techniques. The third understanding was that reactors could also be used to produce energy.

U.S. Reactor Development

Plutonium production during World War II was accomplished in five huge reactors built on the banks of the Columbia River, at Hanford, Washington. While they made no use of the heat generated and were not prototypes of the energy-producing reactors of later generations, much valuable experience was gained. Major accomplishments here were minimizing the effects of radiation upon structural materials, the technique of canning uranium cylinders in aluminum to prevent the cooling water from picking up radioactivity, and recognition of the means of avoiding reactor shutdown due to the inevitable production of neutron-absorbing fission fragments. Plutonium produced in the Hanford reactors was fashioned into the weapons tested at Alamogordo, New Mexico, on July 16, 1945, and exploded over Nagasaki, Japan, on August 9, 1945.

After World War II reactor development was still largely associated with the production of fissionable material. An important decision which determined the direction of future U.S. technology, was to use enriched uranium as reactor fuel. The Hanford reactors had used cylinders of natural uranium, which mean that the U-235, present to the extent of 0.7 percent, provided the neutrons that turned some of the U-238 portion into plutonium. But the Manhattan Project had perfected means of physically separating U-235 from U-238, and indeed the bomb dropped on Hiroshima on August 6, 1945, was constructed of this lighter isotope. If, however, the uranium stock was enriched not to bomb purity of over 90 percent, but only to about two to three percent U-235 content, it would provide excellent reactor fuel, with greater production of neutrons to allow the use of ordinary light water as moderator instead of less neutron-absorbing, but more expensive, carbon or heavy water.

The significance of this decision may be seen in the circumstance that, however controversial the subject of reactor proliferation may be, American models sent to other countries are less likely to be diverted to production of plutonium for weapons because they must rely upon the United States for their next load of enriched fuel. (This situation is changing, as other countries are developing enrichment capabilities. A country could, of course, content itself with the weapons producible from one reactor loading and not care about its future fuel supply.)

The American Atomic Energy Commission (AEC), which took office in 1947, sponsored research on a variety of reactor designs. However, its overriding concern with nuclear weapons precluded a forceful and applied civilian reactor program. Somewhat by default, more by the Cold War pressures of the times, and mostly because of the driving personality of Admiral Hyman Rickover, the direction taken by nuclear power was that of ship propulsion, rather than electricity generation. Conventional submarines had to surface or snorkel for air needed in the battery-charging cycle. A nuclear-powered submarine would not depend on air at all, and this ability to cruise the depths indefinitely was recognized by many of Rickover's colleagues as a revolution in undersea warfare. However, Rickover alone felt that the technology was sufficiently developed to begin construction in the early 1950's, instead of requiring an endless, leisurely series of design studies. His perseverance led to the launching of the *Nautilus* in 1954, a vessel powered by a Westinghouse pressurized-water reactor, and the

Seawolf in 1955, which carried a General Electric sodium-cooled power plant. Over 70 nuclear-powered submarines have subsequently been built by the United States (which now constructs no other types), and many more by other countries.

The advantage of nuclear power for submarines is clear, where avoidance of detection is a paramount concern. For surface ships the question of economics is of equal importance, and the Navy has constructed but a handful of nuclear aircraft carriers, cruisers, and frigates. The nuclear merchant vessel *Savannah* made her maiden voyage in 1962, accompanied by widespread predictions of a new era in low-cost shipping, but reluctance to welcome her into foreign ports because of radiation-leakage fears, and debilitating union disputes undercut her value to such a point that within a few years the ship was placed in mothballs. The Soviet Union has the nuclear ice-breaker *Lenin* operating satisfactorily in its Arctic regions, West Germany uses the *Otto Hahn* as an ore carrier, while the Japanese merchant ship *Mutsu* has been plagued by radiation leaks and blockaded from port by angry fishermen, such that its future is uncertain. The American attempt to power an airplane with a nuclear reactor was abandoned in the early 1960's when no one could conceive a realistic mission for it.

Concurrent with the early submarine reactor work the AEC began to encourage industrial interest in central-station electric power generation. But because the subject was shrouded in government security regulations, those engineers and executives who were allowed a view of it in 1952 and 1953 regarded it with uncertainty. On the one hand, nuclear power might well be economically competitive with fossil fuel plants. On the other, government prohibition of private ownership of reactors discouraged industrial initiatives.

In 1953 the new administration of President Dwight Eisenhower decided to make the sharing (with safeguards) of civilian nuclear information and materials a major facet of foreign policy. The President's "Atoms for Peace" program naturally had domestic impact, for the technology to be shared had yet to be fully developed. Before the end of the year the first "full-scale" electrical power demonstration plant was authorized to be built at Shippingport, Pennsylvania (60,000 kilowatts). In 1954 a law was passed authorizing private ownership of reactor facilities, reserving to the government only possession of the nuclear fuel. These actions and many others showed the determination of the government to encourage industrial development of nuclear power. Since rapid success would be most stimulating to the corporate executives, the Shippingport plant used only an upgraded model of the Navy's *Nautilus* reactor and was the responsibility of none other than Hyman Rickover. Because of his penchant to remain with something that worked well, along with his cancellation of sodium-cooled reactor development, American nuclear industry has been heavily committed to water-cooled designs.

The Shippingport, Pennsylvania, pressurized water reactor (PWR) completed in 1957, four years after its authorization. In 1960 two other reactors, designed expressly for commercial electrical-power production, came on-line—one of them a PWR in Rowe, Massachusetts (Yankee, 140,000 KW), and the other a boiling water reactor (BWR) in Morris, Illinois (Dresden, 180,000 KW). Operating experience from this first generation of about half a dozen reactors proved the technological feasibility of this energy source; in fact, they worked better than expected. The costs, however, were well above those for fossil fuel plants, with construction expenses higher than anticipated, and the future clouded by a decrease in price for coal.

Additional problems confronted utilities in the form of public opposition in the early 1960's: a reactor at Bodega Bay in northern California, which was on a geological fault, was cancelled after an earthquake interrupted construction, while a location within the metropolitan New York area was vetoed by the fearful citizens of that city. Yet in less densely populated areas, and especially in those regions with high fossil fuel costs, nuclear technology still seemed to make economic sense while the dangers of accident were relatively remote.

From 1963 a number of utilities decided to "go nuclear," counting on economies of scale, and encouraged by continuing financial incentives from the federal government as well as by low bids by reactor manufacturers interested at this time more in creating an industry than in immediate profits. Though conceived before this period, the forerunner of the second generation of significantly larger plants was built at San Onofre in southern California. Rated at 428,000 KW, and completed in 1967, its construction signaled not only the arrival of "large-scale" units, but the end of a seven-year period in which no construction licenses for commercial nuclear plants had been issued in the United States.

By the mid-1960's nearly half of all power plant capacity ordered by American utilities was nuclear, with sizes jumping from 500,000 KW to 800,000 KW to 1,100,000 KW, and with all units but one of the PWR and BWR varieties. This commitment coincided with rising coal prices and growing public concern over air pollution from coal-fired plants. It also meshed with increasing comprehension of the concept of a "spaceship earth" with finite resources, stimulated no doubt by the American and Soviet space programs, such that at least one of the fuels of the future clearly was uranium. Nuclear projections for 1980 or the end of the century increased rapidly each year.

Many of the plants ordered since the mid-1960's are now completed. There are 58 nuclear reactors (not including military, experimental, and prototypes) in the United States as of 1976, generating about 9 percent (40 million KW) of the nation's electric power. (For comparison, 20 other countries have a total of about 100 units.) More than 150 other commercial reactors have been planned* or are under construction in America, with an additional capacity of some 170 million KW.

Nuclear Energy: Pro and Con

The future of reactors is far from clear. While some environmentalists laud their contributions to air pollution control, especially in contrast to the emissions of coal-burning power plants, and while the 1973 oil embargo by the OPEC countries encouraged the American government to strive for "energy independence," there is much controversy over whether or not reactors are a viable solution.

Utilities, faced with ever-increasing nuclear plant construction costs and expensive time delays due to inevitable litigation over siting, safety plans, environmental impact reports, etc., and rec-

---

*Editor's note:
These plans have met at least a temporary setback in 1979 with the accident at the Three Mile Island nuclear plant in Pennsylvania, causing widespread public concern and resulting in a governmental study of the engineering and procedural safety precautions in nuclear power installations.

ognizing a lower growth rate than anticipated for electricity consumption, are having second thoughts about the need to "go nuclear," and a number of plants have been cancelled. Even more enervating to the industry than economics, however, has been the opposition of large numbers of citizens who urge legislative action against nuclear plant construction or operation and obviously create a climate of uncertainty unfavorable for investment.

The opposition has questioned the need for fission reactors, stressing fusion, coal, oil, geothermal, tidal, wind, solar, and conservation alternatives, and has also claimed that uranium ore reserves will not last much beyond the year 2000. Industry and government data are viewed as self-serving, with the former AEC (now split into the Nuclear Regulatory Commission and the Department of Energy) especially distrusted as both promoter and regulator of the industry. Above all, safety is the issue. A malfunction in the current variety of reactors cannot result in a nuclear explosion because a critical mass of required purity cannot form. However, should a plumbing leak interrupt the flow of cooling water, and should the emergency system fail to function properly, there is a small but finite possibility that the reactor core will melt, a steam explosion will rupture the containment vessel, and the core's intense radioactivity will be vented and carried downwind. Similarly, an earthquake, airplane crash, or sabotage could release massive amounts of radioactivity. Should breeder reactors be brought to a level of technological and economic viability, and this is yet uncertain, it is possible that an explosive critical mass might accidentally be formed in one. While the probabilities of such accidents are extremely small, their consequences, involving many thousands of people and billions of dollars, are vast. Another cause of concern is the requirement that waste products be stored for many thousands of years without leaking into the water or air. Here the technology seems closer to realization than the identification of geologically inactive sites. Transportation of "hot" wastes and of irradiated fuel rods for reprocessing, by rail or truck, is also viewed with alarm.

Accidents of this sort are but one type of danger seen by the opponents. Of greater credibility and immediacy is the creation of a so-called "plutonium economy." Present reactors manufacture some of this element in their normal operation and then consume a portion; breeders would be designed specifically to make plutonium. Despite the hazards of high radioactivity, plutonium's carcinogenic properties, and the expense involved, governments or other groups willing to go to the trouble can chemically separate the plutonium from uranium, fission fragments, etc., and accumulate a stock of atomic bomb material. It takes only about five kilograms for a Nagasaki-size weapon, and there will be millions of kilograms of this element formed in reactors around the world by the end of the century.* Whether produced as a part of national policy or by terrorist organizations, the proliferation of reactors is taken as tantamount to the proliferation of nuclear weapons. Regardless of international treaties requiring inspection or reprocessing in certain facilities, and despite the best protective efforts, some critics feel that the potential for abrogation of treaties, for surreptitious diversion of the one percent of special nuclear materials expected to be "lost" in the system's bookkeeping, or for outright theft, makes the construction and use of atomic bombs inevitable.

Conclusion

For those who note the increasingly fast pace of life in the 20th century, nuclear reactors provide an appropriate touchstone. Thanks largely to Enrico Fermi's genius, the first such device "went critical" just four years after discovery of the physical phenomenon upon which it was based. Within three more years reactors had been used to manufacture enough of a new element to use that substance in the construction of an awesome new weapon. Only a decade after the end of World War II, and without the intense and massive national commitment that characterized the Manhattan Project, reactors, were powering vessels at sea. And during the next two decades (roughly 1955-1975) civilian reactors went from an enthusiastic vision of a limitless, safe, cheap, and pollution-free source of electrical energy to one regarded as quite likely the opposite. Its advocates recognize the liabilities but believe that no technology is without risks, and that these particular dangers have been and will be well contained. They argue that the health and environmental costs of alternative energy sources are higher, and that the energy is needed for a thriving economy.

* This estimate is found in David R. Inglis, *Nuclear Energy: Its Physics and Its Social Challenge* (Reading, Mass. Addison Wesley, 1973), p. 336.

Reactor opponents assume that disaster in one form or another is inevitable, and that this is a technology we should do without.

Beneath the plethora of technical data both sides produce to support their arguments, there is a value judgment to be made. For this is indeed a major social question and the public is being given (or is taking) a rare opportunity to affect its future. Do the risks outweigh the benefits? What implications would rejection of reactors have upon energy consumption, the economy, the need to control population (if resources are insufficient), the quality of life? Is the technological imperative—if it can be done, it will be done—a desirable philosophy for the United States as it enters its third century, or is a newer attitude—a willingness to do with less—more appropriate in an increasingly interdependent world? Such questions are by no means limited to the controversy over nuclear energy, but reactors do provide, all in a single lifetime, a panorama of the successes, failures, uncertainties, and social impact of the introduction of a new technology.

Enrico Fermi saw little of this development. With the same foresight and courage that led him to abandon atomic physics for nuclear physics in the early 1930's—because he considered that the newly devised quantum mechanics would solve all the former subject's main questions—after World War II he left nuclear physics for high energy particle physics, feeling that this would be the exciting region for future investigation. His death in 1954 left this for others to explore, but superb teacher that he was, his students have been prominent in this next generation.

22

ROBERT H. GODDARD AND THE ORIGINS OF SPACE FLIGHT

Barton C. Hacker

Modern rocketry began with the 20th century in the vision of a handful of men who aspired to space travel. It flourished on the resources of governments largely preoccupied by military concerns. The achievement of space flight required both the exercise of technological imagination and the marshalling of institutional support—that is, both individual creativity and social organization. Robert Hutchings Goddard, an American physicist, was one of the people whose creative vision opened the way to the age of space. His path-breaking experiments were almost entirely his alone. With help from a few trusted assistants and financial backing from several private or semipublic patrons, he designed, tested, and patented almost every feature of the modern rocket. But he never became part of the large-scale social enterprise that ultimately launched mankind into space. Paradoxically, the man who helped create space flight, an archetypical expression of 20th-century corporate technology, was himself more closely allied in spirit and style to the individualistic world of the 19th century. Military-financed research and development programs —large, bureaucratic, amply funded projects aimed at making the rocket a reliable bomb-delivery system—proved decisive, and in these Goddard had no part.

Goddard's Career

Goddard was an experimental physicist—a profession that determined the nature of his achievement. He was born in Worcester, Massachusetts, on October 5, 1882 but spent his youth in Boston, returning with his family to Worcester in 1898. Studious but sickly, the young Goddard did not complete high school until 1904. Four years later he graduated from Worcester Polytechnic Institute. After a year as instructor in physics there, he began graduate studies at Clark University, also in Worcester. Goddard received his doctorate in physics from Clark in 1911, spent a research year at Princeton University, then returned to join the Clark faculty, of which he remained a member until his resignation in 1943. By the time he was appointed full professor in 1920, his rocket researches were already well under way.

Goddard first started thinking about chemical rockets as a means of crossing interplanetary space in 1909. Notebook entries from January 24 to February 2 showed his first efforts to calculate rocket action. Some kind of self-contained reaction device based on Newton's Third Law—for every action there is an equal and opposite reaction—offered the only feasible method of traversing the nearly total vacuum beyond the earth's atmosphere. Chemical rockets met these requirements, although in their existing form they were little more than hollow tubes closed at one end and packed with black powder. Goddard's experiments with such rockets during 1915 and 1916 showed them to be hopelessly inefficient and confirmed what he already suspected—that better means of controlling combustion and better combinations of fuel and oxidizer were needed. His two basic patents for "Rocket Apparatus," issued in July 1914, had specified the use of a combustion chamber and nozzle and suggested the potential of liquid propellants, which, in theory, promised much higher energies than solids.

Goddard had gone about as far as he could on his own, and he was looking for outside support. He found it in the Smithsonian Institution, which in January 1917 granted him $5,000 to further his rocket researches. He scarcely had the money in the bank, however, before the United States entered the First World War.

264

Goddard went to work for the U.S. Army Signal Corps. During 1918 he developed a solid-propellant rocket launched from a hollow tube, intended to serve foot soldiers as portable artillery. It was demonstrated at the Aberdeen Proving Ground in Maryland on November 6, 1918, just five days before the Armistice abruptly halted further work.

Goddard still had the Smithsonian money, however, and Clark University also provided some research funds. By 1921 he was concentrating his efforts on a rocket using liquid propellants. On

Robert H. Goddard stands beside the launching platform for the liquid oxygen rocket he built in 1926 in Auburn, Massachusetts.

March 16, 1926 he achieved the world's first liquid-propelled rocket flight at his Aunt Effie's farm in Auburn, Massachusetts. More work led to further flights, notably one in July 1929, whose spectacular brightness and noise caused a sensation. Reporters and curiosity-seekers forced Goddard to seek a new site for his experiments, which he found in the nearby Camp Devens military reservation.

A more important result was the publicity that brought Goddard's work to the notice of Charles Lindbergh, the world-famous aviator and advisor of the wealthy and influential. Lindbergh's concerns about the future of flight after the air had been mastered seemed to be answered in Goddard's projected conquest of space. A long and fruitful relationship began in November 1929 when they met for the first time. Lindbergh quickly arranged for Goddard to see several potential sponsors, which led first to a grant from the Carnegie Institution of Washington, then to a much larger subsidy from Daniel Guggenheim of New York. Goddard, who had been working on rockets in his spare time, could now set up a full-time experimental program in Roswell, New Mexico, starting in mid-1930.

Between 1930 and 1941 Goddard and a small group of assistants conducted over a hundred proving-stand or static tests and attempted 48 rocket launches, of which 31 produced some kind of flight. The rockets became steadily larger and more sophisticated, reaching dimensions of nearly 22 feet in length, 18 inches in diameter, and a quarter-ton in weight fully loaded with gasoline and liquid oxygen. But Goddard received no further support. During World War II he worked on rocket-assisted takeoff and variable-thrust rocket engines for the Navy and Army Air Forces. He died on August 10, 1945, just as the war was ending. His dream of space travel was unrealized, but the large German V-2 missiles that appeared late in the war made the prospect imminent.

Goddard was by no means an obscure figure in his lifetime but his fame has grown posthumously. Only in 1960 did the United States belatedly acknowledge his priority, in response to an administrative claim filed by Esther Goddard and the Guggenheim Foundation in 1951. The military services were using Goddard patents in missiles then under development. The government awarded a $1,000,000 settlement "for rights to use over 200 of Dr. Goddard's patents which cover basic inventions in the field of rockets, guided missiles and space exploration."*

The Dream of Space Travel

Despite his reticence on the subject, Goddard shared with other pioneers of rocketry the dream of space travel. Whether for reasons of personal shyness and sensitivity, the professional caution of the experimental scientist, or patriotic concern for a potentially valuable military secret, he refrained from pronouncements on his work and its possible applications. The one exception was meteorological research, the rocket's role as a tool for studying the upper atmosphere.

Goddard had supported his request for Smithsonian money in 1916 with a substantial manuscript that he had been working on and adding to since 1913. Eventually published as "A Method of Reaching Extreme Altitudes" in the *Smithsonian Miscellaneous Collections* for 1919, the paper offered a comprehensive statement of rocket theory and a rationale for development based on the value of high-altitude research. Only a modest proposal to explode a rocket-delivered load of magnesium powder on the moon and observe the flash from earth hinted at Goddard's deepest motive, the achievement of space travel. In private, however, he worked hard to build a case for space travel, especially in the confidential reports he filed during the 1920's with the Smithsonian and Clark University.

Goddard's fascination with space travel was first aroused when he encountered science fiction as a boy. The catalyst was H. G. Wells, whose *War of the Worlds* began appearing serially in the *Boston Post* in January 1898,** to be followed by Garrett P. Serviss's *Edison's Conquest of Mars*. The young Goddard was enthralled. Years later, in compiling material for an autobiography, Goddard recalled how the novels "gripped my imagination tremendously. Wells's wonderfully true psychology made the thing very vivid, and possible ways and means of accomplishing the

*Goddard was issued 48 patents during his life, with another 35—in process when he died—granted posthumously. The remaining 131 were issued still later, based on Esther Goddard's gleanings from her husband's notes, sketches, and photos.

**The newspaper version, which appeared daily from January 6, had been altered to make Boston the scene of invasion and retitled "Fighters from Mars, or, The War of the Worlds, in and near Boston."

physical marvels set forth kept me busy thinking." In 1932 he sent Wells a letter of appreciation for *War of the Worlds*. "I was sixteen years old," he wrote, "and the new viewpoints of scientific applications, as well as the compelling realism of the thing, made a deep impression. The spell was complete about a year afterward, and I decided that what might conservatively be called 'high-altitude research' was the most fascinating problem in existence. The spell did not break, and I took up physics as a profession . . . ."

What Goddard referred to as "completing the spell" happened on October 19, 1899. He climbed a cherry tree behind the barn. "It was one of the quiet, colorful afternoons of sheer beauty which we have in October in New England," Goddard recalled in his autobiographical notes, "and as I looked toward the fields at the east, I imagined how wonderful it would be to make some device which had even the *possibility* of ascending to Mars, and how it would look on a small scale, if sent up from the meadow at my feet . . . . I was a different boy when I descended the tree from when I ascended, for existence at last seemed very purposive." For the rest of his life, Goddard celebrated October 19 as "Anniversary Day" and visited the tree if he could.

But what was science fiction and why should it have had so forceful an impact on a boy? The term "science fiction" was not coined until the 1920's, but the brand of literature it labeled was by then a century old, its first landmark being Mary Shelley's *Frankenstein* in 1818. The genre flourished during the closing decades of the 19th century in the works of Jules Verne, Wells, Kurd Lasswitz, and a host of less famous writers. The term itself is something of a misnomer, since technology rather than science has ordinarily been its clearest focus. Science fiction was very much a product of the Industrial Revolution, a literary form that expressed the hopes and anxieties of people caught up in the throes of rapid technological and social change. Science fiction may, in fact, be termed the fantasy of technological change, a cultural device for coping with the unsettling experience of powerlessness in the face of a too quickly altering human environment.

A major focus of that unease during the last three decades of the 19th century was the threat of war, all the more disturbing because proliferating military and naval technology rendered the prospects so uncertain. The nature of future war, especially as influenced by new technology, became not merely a topic of professional interest to military officers. From the 1870's on, first in England, later on the Continent and in America, popular newspapers and magazines fed a growing public concern with lurid stories of invading enemy armies and the havoc they wreaked. In writing *War of the Worlds*, Wells united these concerns with scientific speculation about life on Mars and technological ideas about crossing interplanetary space and created one of the enduring themes of modern science fiction. Wells's skill and insight made his novel something a good deal more significant than the potboilers that typified the war-scare literature, but he knew those works and used them.

Military concerns have been a potent stimulus to the technological imagination in Western civilization for centuries. Leonardo da Vinci's drawings of fanciful weapons are famous, but efforts to imagine a new technology of war can be traced back at least to Guido da Vigevano's work in the 13th century, and perhaps much earlier. Such speculations became heady stuff wrapped in the kind of compelling story Wells provided, and Goddard was not the only boy who succumbed. His testimony could be matched by every major figure in the early history of rocketry, and many later ones as well. Wernher von Braun, for example, recalled how he and other German youths "devoured" Kurd Lasswitz's 1897 novel, *Auf zwei Planeten*, "with curiosity and excitement," and Willy Ley noted how that novel had "preconditioned" a whole generation of Germans, himself included, "to taking space-travel theory seriously." *Auf zwei Planeten*, perhaps needless to say, also featured interplanetary war.

Science fiction became a bridge between realms of theory and the matter-of-fact world. Rockets may have been the only feasible basis for space travel, but at the opening of the 20th century it took a real leap of the imagination to see the potential for journeys between the planets in the erratic black-powder devices known chiefly as playthings and fireworks. Science fiction spurred that leap. It set Goddard to thinking about space travel, and his work on rockets grew directly from his efforts to imagine how space travel might be achieved.

Institutional Support and Military Concerns

Between the dream of space flight and its realization was the hard and expensive task of developing a reliable and efficient rocket. Turning ideas into machines took money and organization —more than any individual was likely to have at his disposal. It certainly took more than Goddard had. How to find the resources needed to reach his self-chosen goal was Goddard's constant concern. In some ways he was surprisingly successful, given the widespread skepticism about space travel that prevailed during his lifetime; in dealing with individuals he often proved to be quite persuasive. Winning the support of large, bureaucratic organizations in government or industry, however, was not so easy for Goddard, and he endured a seemingly endless series of frustrations.

None of this was of much concern at the outset. Theorizing about space travel and rockets was cheap, and even the first rocket experiments demanded no great resources. By September 1916, however, when he first requested Smithsonian support, Goddard was feeling the pinch. ''I have reached the limit of the work I can do single-handed,'' he explained, ''both because of expense, and also because further work will require more than one man's time.'' Only three people were involved in the Smithsonian's decision, and Goddard got what he asked for.

Goddard enjoyed a good deal of success in winning a certain kind of institutional support. Such sponsors as Clark University, the Smithsonian Institution, the Carnegie Institution, and the Guggenheim Foundation practiced an updated version of an old tradition of private or semipublic giving. From such patrons Goddard obtained the few thousands of dollars a year he needed to continue his experiments. Over the course of a quarter-century, the sums involved were far from trivial. Grants to Goddard from all non-military sources between 1917 and 1941 totaled $209,940, of which 90 percent came from the Guggenheims. The editors of Goddard's *Papers* judge this total to be larger ''than was received by any other single scientist for a single project up to the time of World War II.''

The problem Goddard faced, and never really resolved, was that the kind of patrons he found could not guarantee the long-term commitment required for a full-fledged development program. Despite the large total he received over his career, his funding was always on a year-to-year basis with no assurance that money for the next year's work could be found. Almost every spring saw a flurry of letters as Goddard sought to learn if his current patron would provide for the coming academic year. Such uncertainties precluded the kind of long-term financial commitments that might have been required, for example, to construct substantial test facilities. On this basis full-scale rocket development was impossible since it required much greater and more reliable resources.

Organized research and development sponsored by government and industry had emerged as a significant phenomenon during the last half of the 19th century, shaping what has been most characteristic of modern technology. Although the individual inventor has remained important and a surprisingly large number of technological innovations have originated in basement laboratories or their equivalent, development and production have normally demanded large-scale organization. Goddard never managed to find that kind of support although it was not for lack of trying.

Given the obvious military promise of rockets, the military establishment and related industries seemed logical backers for Goddard's work, but he suffered repeated rebuffs. His first effort to find outside support for his research, in fact, had taken the form of a letter to the Secretary of the Navy on July 25, 1914, proposing the development of a rocket-propelled aerial torpedo based on his just-issued patents. The offer was rejected. With help from his friends at the Smithsonian in the crisis atmosphere of World War I, Goddard did get an Army contract, but working on a small solid-propellant rocket was really a diversion from the main task. Despite the drying up of military money after the Armistice, Goddard served as part-time consultant on several rocket projects for the Navy and Army from 1920 through 1923, none of which produced practical results; in any case, all involved solid-propellant rockets, and Goddard had decided that liquid-propelled rockets were the best hope.

Military concerns remained in abeyance through the 1920's but revived and intensified in the following decade. As early as 1931 Goddard cited "national defense" to justify his request for the reissue of his basic 1914 patents; the request was denied. In 1933 when he was casting about for other sources of money during a hiatus in Guggenheim support, Goddard again approached the Navy without success. By 1937 rumors were leaking from Germany of secret work on military rockets, a threat that worried both Goddard and Lindbergh. Contacts between Goddard and the military establishment steadily increased, but his quest for military sponsors remained fruitless. He had no better luck, even with

Lindbergh's aid, when he approached companies like DuPont. Although some individuals were intrigued by Goddard's ideas, their organizations would not be converted to Goddard's cause.

Goddard voiced his long-standing frustration to Lindbergh in 1938. Although "the first practical use for liquid-propelled rockets is likely to be for military purposes," he wrote in a letter on July 23, the military authorities seemed unaccountably reluctant to trust civilian researchers. He noted something that scholars have found often to characterize military responses to new weapons, a resistance to change based on nothing more than habitual ways of doing things. "There is strong skepticism among military men regarding such a new development as this. For example, I recently read a manuscript by an Army officer who looks with disfavor on liquid-propelled rockets because of the difficulty of transporting liquid oxygen, forgetting the objections that were raised in the days of black powder to the thermostatically controlled magazines on board ship for modern smokeless powder."

Goddard may have been too harsh. The military officials who rejected his proposals were acting in good faith on the data they had; few of them shared Goddard's vision of the future. His rockets were costly, complex, hand-crafted machines, longer on promise than performance. He could argue that they would improve, and they did, but fully a third of test flights attempted in New Mexico were failures, and some of the successes were marginal. Even the largest of Goddard's rockets were small for military purposes, their payloads limited to a few pounds. Expense was an important factor during the interwar years when military budgets could seldom be stretched to meet all needs for proven weapons, much less for a risky investment like rockets. In turning down Goddard's proposal in 1933, the Navy noted that rocket devices were impractically expensive and that it would most likely take a long time before their performance matched their promise. That later events were to make cost seem irrelevant and less-than-perfect performance adequate does not impugn the judgment of men who did not, after all, have crystal balls, nor the special incentive that promoted German rocket development—rockets were not among the weapons prohibited to Germany by the Versailles Treaty after World War I. "The United States," as Goddard himself later observed, "had no need for long-range rockets at that time."

Goddard inspects a more advanced rocket in his Roswell, New Mexico, laboratory.

When World War II broke out in Europe, Goddard again sought military backing and again was rebuffed. At a major conference in May 1940 his desire to develop large rockets missiles was ignored. Navy, Army, and Air Corps officials wanted nothing more than a rocket device to assist aircraft in taking off. This was the task that Goddard undertook in late 1941 under concurrent contracts with the Navy and Army Air Forces. It was an important job, as was the work Goddard did on the development of a variable-thrust rocket engine later in the war, but it was far less than he might have done.

As far as Goddard was concerned, the German V-2 missiles that appeared late in World War II were little more than an enlarged version of his own prewar rockets. Although he was a sensitive and humane man, he was thinking more about the venture into space that such large rockets brought closer than he was about the 2,754 Londoners who perished under the bombs those rockets carried. The fear Goddard had shared with Lindbergh about German military rocketry proved to have been well founded. In 1930 the Germany Army initiated a secret project to develop liquid-propellant rocket engines. Although then modest in staff and funding by later standards, the German effort soon surpassed Goddard's lonely and methodical experimental program. Goddard may have been the first to pass each milestone, but the practical development and production of large liquid-propelled rockets were the achievements of the German team. And those achievements, rather than Goddard's, led most directly to the actual accomplishment of space flight before the next decade was over. For all his efforts to make it something more, Goddard's work remained an essentially personal project, and it died with him.

Goddard's Achievement

In evaluating Goddard's contribution to the rise of modern rocketry and the origin of the space age, a clear distinction must be made between invention and theoretical-experimental work, on the one hand, and development and practical application, on the other. Goddard was one of the founders of modern astronautics, playing a vital role in formulating the theoretical basis for space flight, and he devoted his life to creating a liquid-propelled rocket, which he recognized as the means to that end.

In the long run, however, his methodical, one-step-at-a-time approach and his maintenance of rocket work as a personal project veiled in secrecy allowed others to outstrip him. How much of this simply reflected the realities of Goddard's situation and how much was dictated by his character and personality, cannot easily be answered. Whether he was unable to build an organization to carry on his work, or merely unwilling, he pursued a self-directed course that finally left him a lonely and isolated figure.

Salient though military concerns were in the development of rocketry, alive as Goddard was to the military role of rockets, the potential mutuality of interest never became a basis for systematic long-term support. Despite his efforts, Goddard never enjoyed the ample facilities and adequate staffing which allowed his military-backed German rivals to accomplish what he felt he could have done, given the same chance. Western civilization's unique preoccupation with military institutions has meant that military concerns have often furthered enterprise of no obvious military interest. Not only was there a significant military component in the science fiction that first stimulated Goddard and others to work on rockets, but the German expertise acquired in the V-2 development program became the starting point for space flight after World War II. Modified German military technology provided the rockets that launched the first satellites. Only then did Goddard's priority receive a belated acknowledgment.

Elaborate social organization is perhaps more characteristic of engineering than science, and the achievement of space flight is more an engineering than a scientific problem. Goddard the physicist, inspired by his personal vision of space travel, could play a part in building the theory and solving some of the basic problems, but ultimately his work had little direct role in engineering the achievement of space flight itself. That was largely a by-product of the military development of weapons of mass destruction.

23

FREDERICK E. TERMAN AND THE RISE OF SILICON VALLEY

James C. Williams

California's Santa Clara Valley is an alluvial plain at the head of San Francisco Bay. It is bounded on the west by the Pacific's Coastal Range mountains and on the east by the Diablo Range. The two meet and pinch off the valley some 30 miles below the bay. Its northern boundary is marked by Palo Alto, the home of Stanford University, and mud flats that extend into the bay. Until the 1960s the Santa Clara Valley was home to prune, apricot, and cherry orchards and a world-class canning and packing industry. Today the world knows it as "Silicon Valley."

The origins of the electronics industry, which gave Silicon Valley its name, reach back a century to a time when California seemed to be almost a colony of the United States. Geographically isolated from the rest of the nation, the region developed somewhat autonomously. Its principal city, San Francisco, became devoted to shipping and eventually to capital investment. Its major industrial activities were extractive in nature—mining, logging, and farming. It lagged behind the East in developing manufacturing industries partly because Californians had to import the essential industrial energy resource—coal. Yet four decades of gold mining had provided Californians with a rich knowledge of hydraulic engineering. By the 1890s local entrepreneurs and engineers talked of harnessing water power to generate electricity, learning from and even using hydraulic work developed by the mining industry. "By electric transmission," observed one engineer, "[power] can be delivered to those points where it will . . . enable and encourage the establishment of industries, which, without it, would have been impossible."

Regional electric power companies began building hydroelectric plants high in the Sierra Nevada mountains, using the Pelton water wheel, developed for the mining industry in San Francisco, and generators produced by eastern manufacturers. They faced the challenge of pioneering high-tension, long-distance electric power transmission on their own if they were to bring electricity to San Francisco and other coastal cities. By 1901 they had succeeded, and in the process power company engineers and university engineering professors developed a cooperative style of research and development.

Field experience led to invention and innovation, university laboratory experiments, and cooperative testing. For example, in 1898 Stanford professor Frederic A. C. Perrine, his students, and power company engineers successfully field-tested a regionally developed high-potential oil switch which led directly to construction of 40,000 volt lines. Perrine also took a two-year leave from Stanford to consult with the new Standard Electric Company of California, where he coordinated field and laboratory tests of aluminum wire, helping to introduce it for use in California systems. Other Stanford professors involved in similar work included E. E. Farmer and George H. Rowe.

Interchanges between practicing engineers and professors sustained high-tension, long-distance transmission advances. They shared their designs and tests of insulators, switches, and transformers, and employed scientific instruments, such as the oscillograph, to study transmission phenomena. They communicated easily, professors seeking the opinion of "practical men" and speaking to them in an equally practical way. As one power industry representative put it, the professors tore "the mask of ambiguity from electrical theory." They did not lead us "through an interminable labyrinth of mathematics [or] . . . through catacombs of theory—but instead [avoided] these dreary wastes by at once opening up a panorama of results. . . ."

Stanford University's electrical engineering program had become an important element of the California electric-power industry within a decade of its opening. Its electrical engineering graduates found ready work in expanding power companies, and its professors worked closely with the industry. In 1905 Harris J. Ryan, transmission research pioneer at Cornell University, came to Stanford to head the young department. He continued the cooperative university-industry style of research and development, installing the first high-voltage laboratory in the West in 1913 and the first two-million-volt university laboratory in America in 1926.

Meanwhile, shipping interests at the growing port of San Francisco became interested in the wireless communication system developed in Europe by Guglielmo Marconi. Soon after 1900, wireless firms appeared on the Pacific Coast to serve the region with ship-to-shore communications. They also created an intense interest in radio technology among young amateurs, who built wireless sets with Quaker Oats boxes. Historian Arthur Norberg notes that "this field of wireless was ripe for innovation and excellent for exploitation with minimal capital. The only obstacle for amateur and entrepreneur alike was how to circumvent the Marconi patents."

In 1909 Cyril F. Elwell, a recent graduate from Stanford University's electrical engineering program, bypassed Marconi by purchasing exclusive American rights to Danish scientist Valdemar Poulsen's wireless patents. In Palo Alto he demonstrated the system and gained financial backing to organize a company from Stanford University president David Starr Jordan and several professors. Elwell's new firm headquartered itself in San Francisco and quickly expanded. By 1911 a manufacturing branch called the Federal Telegraph Company formed and established a laboratory in Palo Alto near Stanford. Now young university graduates had a backyard opportunity in a new branch of the electrical field.

* * *

Into this fruitful environment in 1910 arrived Frederick E. Terman, a bright and talented boy of ten. His father, psychologist Lewis M. Terman, had joined the Stanford faculty, bringing his

The Federal Telegraph Laboratory, Palo Alto, California.

Source: California History Center Foundation, Cupertino

family from Indiana. The elder Terman already had a reputation as an expert in human intelligence and soon developed the Stanford-Binet intelligence quotient, the basis for standard IQ testing. Meanwhile, Frederick hunted rabbits with a 22-caliber rifle in the hills of the Stanford campus, fished for bass in Felt Lake, and learned to swim in Lake Lagunita. He also caught the wireless craze, and by age 16 the young radio "ham" had built a transmitter with Herbert Hoover, Jr.

Not surprisingly, young Terman pursued a college education at Stanford, graduating in 1920 with a bachelor's degree in chemical engineering. He worked briefly at the Federal Telegraph lab and then enrolled in Harris J. Ryan's electrical engineering program. After graduation in 1922, he was urged by Ryan and his father to attend the Massachusetts Institute of Technology (MIT). Within two years, he earned his doctorate in electrical engineering under the tutelage of Vannevar Bush, and he was offered a teaching position there. Terman returned home in 1924 to prepare himself for his new career in the east, but he was struck down with tuberculosis.

In 1925 Terman's old mentor, Ryan, opened a small radio com-

munications laboratory in the attic of the campus electrical engineering building. Although the young man had still not fully
recovered from his illness, Ryan offered him a half-time teaching
job. Terman had already made his mind up to work in the radio
field and accepted at once. "I took the course in the theory of high
voltage lines," he recalled years later, "and just modified it into a

Frederick E. Terman

*Source: Department of Special Collections and University Archives, Stanford
University Libraries*

generalized theory of long lines. Now this included telephone lines, radio frequency transmission lines, antennas, artificial lines, filters, and so on, as well as high voltage power lines. You could generalize this theory and teach the whole thing in one package. Then the research with students increasingly got to be related to vacuum tubes and circuits associated with vacuum tubes."

Terman's experience at MIT had convinced him that Stanford was falling behind other engineering schools and needed better research with more support to catch up. In 1927, while in charge of the radio laboratory, he wrote an article in *Science* suggesting that engineering research done in corporations was better than that done at universities. He sent a copy to Ryan along with a letter asking for stronger university research support. "With its past reputation as a center of high voltage research, and with the establishment of the Ryan Laboratory," he wrote, "Stanford is in an excellent strategic position to initiate a pioneer movement that will make this the national research center of electrical engineering." Just a modicum of support, he thought, would do wonders for recruiting students and building prestige.

Ryan was swayed. Terman got additional support, recruited bright students, and began ambitious research projects in radio-wave propagation and vacuum tube design. The radio lab soon became the focal point for what Terman described years later as "electronic nuts, those young men who show as much interest in vacuum tubes, transistors, and computers as in girls." He supervised 33 advanced degrees during his first six years, half of the department's total, and after Ryan retired in 1932 more and more electrical engineering students gravitated to Terman. "Studious, soft-spoken, and forever self-effacing," wrote Michael Malone, a regional journalist; Terman's ability to synthesize knowledge "made him first a brilliant teacher and later a profound visionary."

One of the things that attracted students to Terman was his outreach to the local electronics industry. He explained that he "took an interest in what companies there were because there were some jobs there and these were interesting activities anyway. I would take the boys out to see these activities, see what the world off campus was like, and sometimes have people from these companies come in to give talks to the students." This became especially important after Federal Telegraph relocated to the East in 1932, because it had been the prime local employer for early

Stanford communications graduates. Federal's departure left "kind of a semi-desert out here, and one was interested in these [small] companies and helping where they could." Among the remaining firms were Heintz and Kaufmann and Eitel-McCullough in San Francisco and the new Litton Engineering Laboratories in Redwood City. Both Ralph Heintz and Charles Litton were Stanford graduates.

Terman also sought linkages with other university departments. In 1929 he wrote in his annual report on the laboratory: "It is anticipated that from time to time in the future cooperation from the chemistry, physics, and mathematics department will be very desirable and perhaps necessary, in research on certain problems." Such connections were eventually made in large part because of the work of Stanford graduate William W. Hansen, who had joined the physics department in 1934. Hansen was excited by nuclear research and decided to accelerate electrons to probe the nucleus using X-ray tubes and the Ryan High Voltage Laboratory, but lack of money during the Depression years prevented this approach. Instead Hansen pushed forward his research through resonant techniques and soon developed a cavity resonator to accelerate electrons, which he called a "rhumbatron."

At this time a young pilot, Sigurd Varian, persuaded his brother to open a small laboratory near their home in Halcyon, a town 225 miles south of Palo Alto. Sigurd was interested particularly in developing an aircraft navigation and detection system, and his brother Russell was a Stanford physics graduate who coincidentally had been William Hansen's roommate when they were both students. Russell wrote Hansen with an idea that the rhumbatron might be useful in Sigurd's project. Hansen agreed, and a collaboration began that brought the brothers from Halcyon to Stanford. The university agreed to take them on as unpaid research associates with $100 a year in materials and laboratory space in return for half-interest in any resulting patents. Within a few months they had developed a new electron tube dubbed the klystron, which seemed to be the key to the navigation and detection problem. They quickly sought and received financial support from the Sperry Gyroscope Company in New York, a connection that launched increased microwave research at Stanford.

Meanwhile, Terman was named head of Stanford's Department of Electrical Engineering in 1937. Two of his students, William

Hewlett and David Packard, had met on campus and then gone separate ways after graduating in 1934. Packard took a job in New York with General Electric, and Hewlett continued with Terman for a while and then went to MIT for further study. In 1936 Terman had helped Hewlett get a job and also brought Packard back to Palo Alto with a graduate fellowship. The fellowship came indirectly from Hansen's work with the Varian brothers and the Sperry Company. Charles Litton had patented some multigrid tubes in which oscillations occurred and assigned the patents to Stanford. They were put into the same pool as the klystron patents. Arthur Norberg explains what happened when Sperry licensed the pool of patents: "Stanford received an extra $1,000 for the Litton idea and Litton insisted the money go to the engineering school for research on electronic tubes. Terman allocated it to graduate student support, specifically for David Packard."

In 1937 Terman encouraged Hewlett and Packard to form a business to commercialize the audio-oscillator developed by Hewlett under his mentor's guidance. In their tiny shop located in Packard's garage, Hewlett perfected the audio-oscillator as a distortion analyzer which generated signals of different frequencies. They also produced an electronic frequency meter, and Terman was able to tell when they had orders for their devices: "If the car was in the garage there was no backlog. But if the car was parked in the driveway, business was good." By 1940 the two young engineers' growing business had nine employees and would expand rapidly during the war years.

In addition to helping Hewlett and Packard, Terman had also directed several students to William Hansen's klystron project. On the eve of World War II, they were successful in making the klystron a practical microwave-radio device with many applications. The war brought a variety of opportunities and changes. The Sperry Company became an important war contractor, and Terman's connection to Hansen's work helped him land $10,000 from Sperry for klystron research in Stanford's communications program. Meanwhile, Sperry relocated Hansen and his physics team to its Long Island research center for the duration. Before leaving in December 1940, however, Hansen taught a special class on the klystron for Terman's graduate students, preparing them for the war work they would be doing with Sperry's support. As historians Stuart Leslie and Bruce Hevly described the impact,

"Sperry's support of klystron research back at Stanford trained a new generation of microwave engineers. . . . In just a few years Stanford's electrical engineering program had gone from providing a couple of graduate students for the physics department microwave studies to a full-fledged research commitment of its own."

Terman himself was affected profoundly by the war. In 1942 Vannevar Bush, Chair of the Office of Scientific Research and Development, asked him to move to Boston to take over the top-secret Radio Research Laboratory at Harvard University. There he managed a research program involving over 800 people, devised jamming devices for radar, and developed tunable receivers for detecting and analyzing radar signals. He lived across the street from Harvard University's treasurer, and because Terman's project brought in over half the money Harvard was spending on war research, the two became friendly. On Sundays Terman used to chat with his Harvard acquaintance, following him around as he worked in his garden. "I asked him what he thought would happen after the war. It seemed to me there'd be a new wave of government support. The scientific war effort had been so very successful."

Terman's views were reinforced, and although there had been little peacetime government support of university research before the war, he became convinced this would change. "The war had made it obvious to me that science and technology are more important to national defense than masses of men. The war also showed how essential the electron was to our type of civilization." After the war "there would be, for the first time, real money available to support engineering research and graduate students. This new ballgame would be called sponsored research," he said.

* * *

Frederick Terman observed that "Stanford emerged from World War II as an underprivileged institution. It had not been significantly involved in any of the exciting engineering and scientific activities associated with the war." When he returned in 1946 and assumed leadership as dean of engineering, he moved forward with the view that Stanford could achieve a position of national importance in electronics. It already had a strong reputation in high-voltage power transmission and, he noted, "had made some very important contributions under Harris J. Ryan." But the "newer opportunities, the opportunities of the future were in electronics and things related to electronics." So Terman launched a simple plan to attract the brightest faculty who would seek out support for the most desirable research projects to attract the best graduate students. At the same time, they would develop close connections with private industry. He dubbed his approach "steeple building," and since Stanford already had an edge in microwave technology because of its klystron work, he looked to it as the foundation for the future.

"After the first month or two I was back at Stanford," recalled Terman, "a couple of boys showed up from the Navy. The Office of Naval Research (ONR) had just been established and they headed for my lab first." Terman talked over some ideas with the Navy, got university president Donald Tresidder to back him, and landed a $225,000 annual contract for basic research. "We started off with three projects. The one in chemistry fizzled. The physics project led to the Nobel prize for Dr. Felix Bloch, who discovered nuclear magnetic resonance. The project in electrical engineering was a seed that blossomed into today's nationally recognized research program in engineering." This initial start in sponsored research led to Stanford's Electronics Research Laboratories (ERL) as well as to a project for the development of the Stanford Linear Accelerator Center.

The ERL carried out fundamental studies in electronics. While researchers were not doing applied research and bristled at the idea the ERL might become a development laboratory for military or commercial ventures, they did follow Terman's lead in building ties outside the university. After the war, Russell and Sigurd Varian had opened a business in San Carlos, not far from Palo Alto, joining Hewlett-Packard, Litton Engineering, and other home-grown electronics firms. ERL researchers worked informally and sometimes formally with these and other companies converting their discoveries into practical hardware; the number of innovations coming from ERL's fundamental research gave Stanford a position of increasing importance to military and commercial interests. It is therefore not surprising that the ONR came first to Terman with an applied research proposal after the Korean War began in 1950; Terman got university support for it within two weeks.

The enormous $450,000 annual applied electronics research contract consolidated Stanford's position in electronics. Terman became director of the new Applied Electronics Laboratory (AEL) housed in a new facility built with ONR dollars and a gift from Hewlett-Packard. In directing the program, he accepted only projects that strengthened Stanford's basic electronics research and enhanced the university's reputation. He also redoubled his efforts to build stronger ties with industry. He had been doing this for years, speaking to industrial groups about the benefits of their being acquainted with university research work and getting to know the fellows in local industries. But now this "community of interest between the University and local industry" became essential, since the AEL was to produce prototype electronic devices and then work with the firms which actually would produce them.

A spur to this "community of interest" came from Stanford's own financial problems at the time. The university's endowment was too small to provide either adequate operating or capital improvement funds, and its founding grant prevented it from selling off any of its 8,100 acres of land. Administrators, working after 1951 with a faculty committee, concocted several plans for generating income from the land, including agricultural development, a regional shopping center, and housing projects. A 40-acre corner was earmarked for a little light industry. Most of the planners' energy went into the shopping center and housing, and a master plan would not be finalized until 1953. But Varian Associates looked at the industrial land and took the first initiative toward developing it.

Varian Associates, according to historian Henry Lowood, specialized in developing products from research conducted at Stanford and regularly hired university faculty, research associates, and students. It was so successful that by 1949 it had outgrown its San Carlos facilities. Russell Varian and Edward Ginzton, whom Terman had steered to the klystron project as a student before the war and who now was a Varian director as well as a Stanford professor, decided to build a branch facility near the university. Early in 1950, Varian asked Stanford to lease land to the firm for construction of a new research and development facility. The agreement seemed eminently sensible. The firm's research was based on university patents, and Stanford faculty numbered among its important shareholders, directors, and consultants.

This move led Terman, also a Varian director, to give steerage to the university's developing land-use master plan. He saw the opportunity to lease only to companies with activities connected to the university's programs, yet the master plan seemed to include other industries, even insurance companies. The second lease to Eastman Kodak for a photo-processing plant mobilized Terman. By 1953 he was on the university's Advisory Committee on Land and Building Development and busy persuading Alf Brandin, the university business manager, to come around to his way of thinking.

Terman pulled his arguments from several directions. He restated that the park could reinforce the university's steeples of excellence in research. He pointed out the large gifts Stanford had received from Hewlett-Packard, Varian Associates, and Russell Varian. These companies were good donors, their gifts equalling the university's lease income. He also pointed out the important matching tuition paid by local firms for their employees who participated in the honors cooperative program which enabled outside engineers to sit in on Stanford classes directly or via a television network. "Brandin was quick on the trigger," he recalled. "Very soon thereafter, if you weren't a high-technology company, you had a hell of a time coaxing him to give you a lease."

In 1955 Terman became the university provost and turned to building up Stanford's chemistry department as the foundation for a community of interest in biotechnology. Soon he could see his work bearing fruit. His engineering steeples of excellence made Stanford one of the nation's foremost research universities. Sponsored research climbed from the first ONR project of $225,000 in 1946 to $12 million in 1967. This stimulated even more growth; corporate gifts reached a half a million dollars in 1956 and rose to over $2 million by the year 1965. Meanwhile, Hewlett-Packard became the flagship of Stanford Industrial Park in 1956, and soon Ampex and Lockheed's new Space and Missile Division moved in. By 1960 over 40 firms occupied the 450-acre campus-style park, and Terman was being asked to help mastermind similar feats in Texas and New Jersey.

* * *

The Santa Clara Valley, once known by local agricultural boosters as "the Valley of Heart's Delight," had become a growing center for the electronics and aerospace industries. Defense contracts helped drive the boom, the earliest going to Stanford and its spreading community of electronics companies. Development of the NASA-Ames Research Center (opened in 1940) in Sunnyvale also brought contracts and stimulated industry, and San Jose Chamber of Commerce manager, Russell Petit, spurred the region's growth with an intensive national advertising campaign to attract even more companies. Cashing in on the vacuum tube, radar, microwave technology, the cold war, and space exploration, giant national firms of the tube era of electronics as well as young new entrepreneurs set up shop from Palo Alto south to San Jose. The spreading high-technology community attracted national attention. "Any electrical engineer with ambition ran his resume around the Valley circuit in hope of a response," noted Michael Malone, and seasoned researchers came West as well.

In 1956 William B. Shockley returned to his boyhood home in Palo Alto. As coinventor of the transistor while at Bell Telephone Laboratories in 1947, he had just won the Nobel Prize with John Bardeen and Walter Brattain. Now Shockley would establish the first semiconductor company in the fertile valley. The very brightest young engineers in the country responded to his call, and he hired the best. But within two years, his young protégés brewed a rebellion. Disagreeing with Shockley's research directions and fed up with his contemptuous treatment of them, seven began looking for support to set up on their own. With almost no venture capital resources on the West Coast, a New York investment firm finally put them in touch with Fairchild Camera and Instrument Corporation in New Jersey. Fairchild seemed interested but was concerned that none of the seven possessed clear management skills. To win the company's support, the seven recruited the only holdout at Shockley who possessed the needed leadership, 27-year-old Robert Noyce. The now eight men quit en masse before

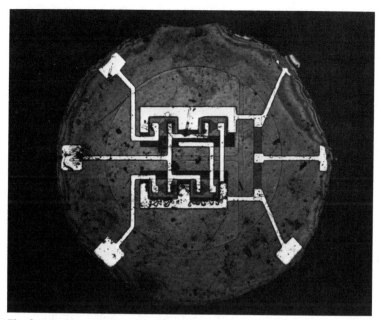

The first mass-producible monolithic integrated circuit, developed by Fairchild Semiconductor in 1961.

Source: California History Center Foundation, Cupertino

an enraged Shockley. Fairchild Semiconductor opened in Mountain View, and Silicon Valley was born.

Numerous problems plagued Fairchild from the start, even though it grew rapidly and introduced the first mass-producible monolithic integrated circuit. Michael Malone described it as "a corporate vocational school" for young engineers. "Here they could screw up without serious repercussions—after all, nobody else knew how the job was done either—and learn from their mistakes." They did, unleashing the centrifugal forces of entrepreneurial creativity which have since characterized Silicon Valley. By 1968 the last of the original eight founders, along with many other Fairchild engineers, had moved on to replicate the founders' experiences in new ventures. Later Robert Noyce remarked that it came "as a great revelation—and a great motivation, too," that a young engineer or scientist could get venture capital for a new company. At a 1969 conference in Sunnyvale, fewer than two dozen of the 400 semiconductor engineers present had not worked for Fairchild, and by the early 1970s, former Fairchild employees had started 41 new semiconductor companies, many of them in the valley.

The only apparent early connection between Fairchild and the Stanford community which Terman had built rested in the firm's early hiring of graduate students as production workers. A closer tie did not come until Marcian "Ted" Hoff joined Noyce's new Intel Corporation after receiving his Ph.D. in electrical engineering at Stanford. In 1969 Hoff developed the microprocessor, the first computer on a chip. Intel's 4004 chip launched another explosive wave through the valley and beyond. By 1975 Intel's third microprocessor, the 8080, became the heart of the first affordable computer, the Altair, produced in Albuquerque, New Mexico.

Popular Electronics' announcement of the Altair in January 1975 brought into existence Silicon Valley's Homebrew Computer Club. Its first meeting in March drew 32 people to a home in Menlo Park, near Palo Alto. Within weeks the meetings were drawing several hundred enthusiasts, and it began meeting at the Stanford Linear Accelerator Center auditorium. According to authors Paul Freiberger and Michael Swaine, the Homebrew Computer Club provided "the intellectual nutrient" in which the valley's microcomputer companies first swam. Homebrewers "thrived in a kind of joyous anarchy" from which solid engineering and technologies for people emerged.

Stephen Wozniak, a young employee at Hewlett-Packard and gifted computer hobbyist, was a Homebrewer from the start. He went regularly, absorbing the home-built machines others brought to the meetings and beginning to feel he could improve on their designs. He bought one of the latest microprocessor chips in conjunction with a San Francisco computer show and built a computer of his own. Wozniak took it to Homebrew, showed it off, and passed out photocopies of his design for the Apple. Meanwhile, his friend Steven Jobs suggested they start a company, and in 1976 Apple Computer was born in the tradition of Hewlett and Packard—in a garage in Cupertino. Eight years later its sales topped $1.5 billion, and Silicon Valley was the envy of the world.

Frederick E. Terman, who had retired from Stanford in 1965, must have been pleased with this latest odyssey in the Santa Clara Valley. After all, it was his vision and life's work that had created the foundation from which it sprang. In 1952 he had observed, "Almost anything that one wishes to do in the world of today is made possible, or is done better, or is helped by electronics. Through its ability to control, to amplify, and to convert between light, sound, and electricity, electronics provides a nervous system for our machine-age civilization." By the time of his death in 1982, his dreams had been more than fulfilled.

FOR FURTHER READING

1 TECHNOLOGY IN AMERICA: AN INTRODUCTION

Hindle, Brooke, and Steven Lubar. *Engines of Change: The American Industrial Revolution, 1790–1860*. Washington, DC, 1986.

Hughes, Thomas P. *American Genesis: A Century of Technological Enthusiasm, 1870–1970*. New York, 1989.

Marx, Leo. *The Pilot and the Passenger: Essays on Literature, Technology, and Culture in the United States*. New York, 1988.

Morison, Elting E. *From Know-how to Nowhere: The Development of American Technology*. New York, 1974.

Smith, Merritt Roe, ed. *Military Enterprise and Technological Change: Perspectives on the American Experience*. Cambridge, MA, 1985.

2 THE ARTISAN DURING AMERICA'S WOODEN AGE

Bedini, Silvio A. *Thinkers and Tinkers: Early American Men of Science*. New York, 1975.

Hindle, Brooke, ed. *America's Wooden Age: Aspects of its Early Technology*. Tarrytown, NY, 1975.

Hindle, Brooke. *Technology in Early America: Needs and Opportunities for Study*. Chapel Hill, 1965.

McPhee, John. *The Survival of the Bark Canoe*. New York, 1975.

Sloane, Eric. *A Reverence for Wood*. New York, 1965.

Sloane, Eric. *Our Vanishing Landscape*. New York, 1955.

3 THOMAS JEFFERSON AND A DEMOCRATIC TECHNOLOGY

Ferguson, Eugene S. *Oliver Evans: Inventive Genius of the American Industrial Revolution*. Greenville, 1980.

Hindle, Brooke. *Emulation and Invention*. New York, 1981.

Kasson, John. *Civilizing the Machine: Technology and Republican Values in America*. New York, 1976.

Kouwenhoven, John. *Made in America*. New York, 1948.

Marx, Leo. *The Machine in the Garden: Technology and the Pastoral Ideal in America*. New York, 1964.

292

4 BENJAMIN HENRY LATROBE AND THE TRANSFER OF TECHNOLOGY

Jeremy, David J. *Transatlantic Industrial Revolution: The Diffusion of Textile Technologies between Britain and America, 1790–1830*. Cambridge, MA, 1981.

Pursell, Carroll W., Jr. *Early Stationary Steam Engines in America: A Study in the Migration of a Technology*. Washington, DC, 1969.

Stapleton, Darwin H. *The Transfer of Early Industrial Technologies to America*. Philadelphia, 1987.

Tucker, Barbara M. *Samuel Slater and the Origins of the American Textile Industry, 1790–1860*. Ithaca, 1984.

5 ELI WHITNEY AND THE AMERICAN SYSTEM OF MANUFACTURING

Hoke, Donald. *Ingenious Yankees: The Rise of the American System of Manufactures in the Private Sector*. New York, 1989.

Hounshell, David A. *From the American System to Mass Production, 1800–1932: The Development of Manufacturing Technology in the United States*. Baltimore, 1984.

Mayr, Otto and Robert C. Post, eds. *Yankee Enterprise: The Rise of the American System of Manufactures*. Washington, DC, 1981.

McGaw, Judith A. *Most Wonderful Machine: Mechanization and Social Change in Berkshire Paper Making, 1801–1885*. Princeton, 1987.

Noble, David F. *Forces of Production: A Social History of Industrial Automation*. New York, 1984.

Rosenberg, Nathan, ed. *The American System of Manufactures*. Edinburgh, 1969.

6 THOMAS P. JONES AND THE EVOLUTION OF TECHNICAL EDUCATION

Calvert, Monte A. *The Mechanical Engineer in America, 1830–1910: Professional Cultures in Conflict*. Baltimore, 1967.

Ferguson, Eugene S., ed. *Early Engineering Reminiscences (1815–40) of George Escol Sellers*. Washington, DC, 1965.

Post, Robert Charles. *Physics, Patents and Politics: A Biography of Charles Grafton Page*. New York, 1976.

Sinclair, Bruce. *Philadelphia's Philosopher Mechanics: A History of the Franklin Institute, 1824–1865*. Baltimore, 1974.

7 CYRUS HALL McCORMICK AND THE MECHANIZATION OF AGRICULTURE

Danhof, Clarence H. *Change in Agriculture: The Northern United States, 1820–1870*. Cambridge, MA, 1969.

Hutchinson, William T. *Cyrus Hall McCormick*. 2 vols., New York, 1930–35.

Rossiter, Margaret W. *The Emergence of Agricultural Science: Justus Liebig and the Americans, 1840–1880*. New Haven, 1975.

Wik, Reynold M. *Steam Power on the American Farm*. Philadelphia, 1953.

Williams, Robert C. *Fordson, Farmall, and Poppin' Johnny: A History of the Farm Tractor and Its Impact on America*. Urbana, IL, 1987.

8 JAMES BUCHANAN EADS AND THE ENGINEER AS ENTREPRENEUR

Chandler, Alfred D. *The Visible Hand: Managerial Revolution in American Business*. Cambridge, MA, 1977.

Hughes, Thomas Parke. *Elmer Sperry: Inventor and Engineer*. Baltimore, 1971.

McCartney, Laton. *Friends in High Places. The Bechtel Story: The Most Secret Corporation and How It Engineered the World*. New York, 1988.

McMahon, A. Michal. *The Making of A Profession: A Century of Electrical Engineering in America*. New York, 1984.

Sinclair, Bruce. *A Centennial History of the American Society of Mechanical Engineers, 1880–1980*. Toronto, 1980.

Merritt, Raymond H. *Engineering in American Society, 1850–1875*. Lexington, 1969.

9 JAMES B. FRANCIS AND THE RISE OF SCIENTIFIC TECHNOLOGY

Hunter, Louis C. *A History of Industrial Power in the United States, 1780–1930. Volume One: Waterpower in the Century of the Steam Engine*. Charlottesville, 1979.

Hunter, Louis C. *A History of Industrial Power in the United States, 1780–1930: Volume Two: Steam Power*. Charlottesville, 1985.

McHugh, Jeanne. *Alexander Holley and the Makers of Steel*. Baltimore, 1980.

Noble, David F. *America by Design: Science, Technology, and the Rise of Corporate Capitalism*. New York, 1977.

Spence, Clark C. *Mining Engineers & The American West: The Lace-Boot Brigade, 1849–1933*. New Haven, 1970.

10 ALEXANDER GRAHAM BELL AND THE CONQUEST OF SOLITUDE

Bruce, Robert V. *Bell: Alexander Graham Bell and the Conquest of Solitude*. Boston, 1973.

Smith, George David. *The Anatomy of a Business Strategy: Bell, Western Electric, and the Origins of the American Telephone Industry*. Baltimore, 1985.

Thompson, Robert Luther. *Wiring a Continent: The History of the Telegraph Industry in the United States*. Princeton, 1947.

Wasserman, Neil H. *From Invention to Innovation: Long-Distance Telephone Transmission at the Turn of the Century*. Baltimore, 1985.

11 THOMAS ALVA EDISON AND THE RISE OF ELECTRICITY

Friedel, Robert and Paul Israel. *Edison's Electric Light: Biography of an Invention*. New Brunswick, 1986.

Josephson, Matthew. *Edison: A Biography*. New York, 1959.

MacLaren, Malcolm. *The Rise of the Electrical Industry During the Nineteenth Century*. Princeton, 1943.

Passer, Harold C. *The Electrical Manufacturers, 1875–1900: A Study in Competition, Entrepreneurship, Technical Change, and Economic Growth*. Cambridge, MA, 1953.

Wachhorst, Wyn. *Thomas Alva Edison: An American Myth*. Cambridge, MA, 1981.

12 GEORGE EASTMAN AND THE COMING OF INDUSTRIAL RESEARCH

Hounshell, David A., and John Kenly Smith, Jr. *Science and Corporate Strategy: Du Pont R&D, 1902–1980*. Cambridge, MA, 1988.

Jenkins, Reese V. *Images & Enterprise: Technology and the American Photographic Industry, 1839–1925*. Baltimore, 1975.

Jewkes, John, David Sawers, and Richard Stillerman. *The Sources of Invention*. London, 1958.

Reich, Leonard S. *The Making of American Industrial Research: Science and Business at GE and Bell, 1876–1926*. Cambridge, MA, 1985.

Wise, George. *Willis R. Whitney, General Electric, and the Origins of U.S. Industrial Research*. New York, 1985.

13 ELLEN SWALLOW RICHARDS: TECHNOLOGY AND WOMEN

Cowan, Ruth Schwartz. *More Work for Mother: The Ironies of Household Technology from the Open Hearth to the Microwave*. New York, 1983.

Dublin, Thomas. *Women at Work: The Transformation of Work and Community in Lowell, Massachusetts 1826–1860*. New York, 1979.

Hayden, Dolores. *The Grand Domestic Revolution: A History of Feminist Designs for American Homes, Neighborhoods, and Cities*. Cambridge, MA, 1981.

Rothman, Barbara Katz. *Recreating Motherhood: Ideology and Technology in a Patriarchal Society*. New York, 1989.

Strasser, Susan. *Never Done: A History of American Housework*. New York, 1982.

14 GIFFORD PINCHOT AND THE CONSERVATION MOVEMENT

Hays, Samuel P. *Beauty, Health, and Permanence: Environmental Politics in the United States, 1955–1985*. Cambridge, MA, 1987.

Hays, Samuel P. *Conservation and the Gospel of Efficiency: The Progressive Conservation Movement, 1890–1920*. Cambridge, MA, 1959.

Melosi, Martin V., ed. *Pollution and Reform in American Cities, 1870–1930*. Austin, 1980.

Tarr, Joel A., and Gabriel Dupuy, eds. *Technology and the Rise of the Networked City in Europe and America*. Philadelphia, 1989.

Worster, Donald. *Rivers of Empire: Water, Aridity and the Growth of the American West*. New York, 1985.

15 FREDERICK WINSLOW TAYLOR AND SCIENTIFIC MANAGEMENT

Aitken, Hugh G. J. *Scientific Management in Action: Taylorism at Watertown Arsenal, 1908–1915*. Cambridge, MA, 1960.

Braverman, Harry. *Labor and Monopoly Capital: The Degradation of Work in the Twentieth Century*. New York, 1974.

Copley, Frank Barkley. *Frederick W. Taylor: Father of Scientific Management*. 2 vols., New York, 1923.

Gilbreth, Frank B., Jr. *Time Out for Happiness*. New York, 1970.

Kakar, Sudhir. *Frederick Taylor: A Study in Personality and Innovation*. Cambridge, MA, 1970.

Nelson, Daniel. *Frederick W. Taylor and the Rise of Scientific Management*. Madison, 1980.

16 HENRY FORD AND THE TRIUMPH OF THE AUTOMOBILE

Flink, James J. *The Automobile Age*. Cambridge, MA, 1988.

Flink, James J. *The Car Culture*. Cambridge, MA, 1975.

Foster, Mark S. *From Streetcar to Superhighway: American City Planners and Urban Transportation, 1900–1940*. Philadelphia, 1981.

Leslie, Stuart W. *Boss Kettering: Wizard of General Motors*. New York, 1983.

Rose, Mark H. *Interstate: Express Highway Politics, 1941–1956*. Lawrence, 1979.

Seely, Bruce E. *Building the American Highway System: Engineers as Policy Makers*. Philadelphia, 1987.

17 PETER L. JENSEN AND THE AMPLIFICATION OF SOUND

Aitken, Hugh G. J. *The Continuous Wave: Technology and American Radio, 1900–1932*. Princeton, 1985.

Aitken, Hugh G. J. *Syntony and Spark: The Origins of Radio*. Princeton, 1976.

Czitrom, Daniel J. *Media and the American Mind: From Morse to McLuhan*. Chapel Hill, 1982.

Douglas, Susan J. *Inventing American Broadcasting, 1899–1922*. Baltimore, 1987.

Jensen, Peter L. *The Great Voice*. Richardson, TX, 1975.

18 CHARLES A. LINDBERGH: HIS FLIGHT AND THE AMERICAN IDEAL

Bilstein, Roger E. *Flight in America, 1900–1983: From the Wrights to the Astronauts*. Baltimore, 1984.

Bilstein, Roger E. *Flight Patterns: Trends of Aeronautical Development in the United States, 1918–1929*. Athens, 1983.

Constant, Edward W., II. *The Origins of the Turbojet Revolution*. Baltimore, 1980.

Corn, Joseph J. *The Winged Gospel: America's Romance with Aviation, 1900–1950*. New York, 1983.

Rae, John B. *Climb to Greatness: The American Aircraft Industry, 1920–1960*. Cambridge, MA, 1968.

Roland, Alex. *Model Research: The National Advisory Committee for Aeronautics, 1915–1958*. 2 vols., Washington, DC, 1985.

19 KEATON AND CHAPLIN: THE SILENT FILM'S RESPONSE TO TECHNOLOGY

Corn, Joseph J., ed. *Imagining Tomorrow: History, Technology, and the American Future*. Cambridge, MA, 1986.

Kasson, John F. *Amusing the Million: Coney Island at the Turn of the Century*. New York, 1978.

Marzio, Peter C. *Rube Goldberg: His Life and Work*. New York, 1973.

Meikle, Jeffrey L. *Twentieth Century Limited: Industrial Design in America, 1925–1939*. Philadelphia, 1979.

20 MORRIS L. COOKE AND ENERGY FOR AMERICA

Butti, Ken and John Perlin. *A Golden Thread: 2500 Years of Solar Architecture and Technology*. Palo Alto, 1980.

Christie, Jean. *Morris Llewellyn Cooke: Progressive Engineer*. New York, 1983.

Melosi, Martin V. *Coping with Abundance: Energy and Environment in Industrial America*. New York, 1985.

Hughes, Thomas P. *Networks of Power: Electrification in Western Society, 1880–1930*. Baltimore, 1983.

21 ENRICO FERMI AND THE DEVELOPMENT OF NUCLEAR POWER

Badash, Lawrence. *Radioactivity in America: Growth and Decay of a Science*. Baltimore, 1979.

Boyer, Paul. *By the Bomb's Early Light: American Thought and Culture at the Dawn of the Atomic Age*. New York, 1985.

Hewlett, Richard G. and Oscar E. Anderson, Jr. *The New World, 1939/1946: Volume I of A History of The United States Atomic Energy Commission*. University Park, 1962.

Hewlett, Richard G. and Francis Duncan. *Nuclear Navy, 1946–1962*. Chicago, 1974.

Nelkin, Dorothy. *Nuclear Power and Its Critics; The Cayuga Lake Controversy*. Ithaca, 1971.

York, Herbert F. *The Advisers: Oppenheimer, Teller, and the Superbomb*. San Francisco, 1976.

22 ROBERT H. GODDARD AND THE ORIGINS OF SPACE FLIGHT

Lasby, Clarence G. *Project Paperclip: German Scientists and the Cold War*. New York, 1971.

Koppes, Clayton R. *JPL and the American Space Program: A History of the Jet Propulsion Laboratory*. New Haven, 1982.

Lehman, Milton. *This High Man: The Life of Robert H. Goddard*. New York, 1963.

Mazlish, Bruce, ed. *The Railroad and the Space Program: An Exploration in Historical Analogy*. Cambridge, MA, 1965.

McDougall, Walter A. . . . *the Heavens and the Earth: A Political History of the Space Age*. New York, 1985.

23 FREDERICK E. TERMAN AND THE RISE OF SILICON VALLEY

Austrian, Geoffrey D. *Herman Hollerith: Forgotten Giant of Information Processing*. New York, 1982.

Dickson, David. *The New Politics of Science*. 2 ed., Chicago, 1988.

Hanson, Dirk. *The New Alchemists: Silicon Valley and the Microelectronics Revolution*. Boston, 1982.

Kenney, Martin. *Biotechnology: The University-Industrial Complex*. New Haven, 1986.

Kidder, Tracy. *The Soul of a New Machine*. Boston, 1981.

Turkle, Sherry. *The Second Self: Computers and the Human Spirit*. New York, 1984.

Young, Jeffrey S. *Steve Jobs: The Journey Is the Reward*. Glenview, 1988.

CONTRIBUTORS

Lawrence Badash is Professor of the History of Science at the University of California, Santa Barbara. His research deals primarily with the history of nuclear physics and its applications.

George Basalla is Professor of History at the University of Delaware. He specializes in the social history of technology—more particularly how technology relates to culture—and is the author of *The Evolution of Technology* (1989).

Robert V. Bruce is Professor of History at Boston University. In addition to his biography of Alexander Graham Bell, he is author of *Lincoln and the Tools of War* (1956) and the Pulitzer Prize-winning *The Launching of Modern American Science* (1988).

Jean Christie is Professor of History at Fairleigh Dickinson University. Besides her monograph on Morris L. Cooke she has published two collections of essays on twentieth-century American history.

Gail Cooper is Assistant Professor of History at Lehigh University. She is currently working on the adoption of statistical quality control in Japan.

Ruth Schwartz Cowan is Professor of History at the State University of New York at Stony Brook. She is the author of *More Work for Mother* (1983) and many articles on the relationship between women and technology.

James J. Flink is Professor of Comparative Cultures at the University of California, Irvine. He has written three books on the automobile.

Barton C. Hacker teaches history at Oregon State University. He is the author, most recently, of *The Dragon's Tail: Radiation Safety in the Manhattan Project, 1942–1946* (1987).

Samuel P. Hays is Professor of History at the University of Pittsburgh. He is most recently the author of *Beauty, Health, and Permanence: Environmental Politics in the United States, 1955–1985* (1987).

Brooke Hindle retired as Senior Historian at the National Museum of American History of the Smithsonian Institution. He is the author of many books on American technology, most recently (with Steven Lubar),

Engines of Change: The American Industrial Revolution, 1790–1860 (1986).

Thomas Parke Hughes is Professor of the History of Technology at the University of Pennsylvania. His most recent book is *American Genesis: A Century of Technological Enthusiasm, 1870–1970* (1989).

Reese V. Jenkins teaches at Rutgers University and is the director-in-chief of the Thomas A. Edison Papers. He is the author of *Images and Enterprise: Technology and the American Photographic Industry* (1975).

John A. Kouwenhoven is Professor Emeritus of English at Barnard College. His books include the classic *Made in America: The Arts in Modern American Civilization* (1948).

Edwin T. Layton, Jr. is Professor of History, Science, and Technology at the University of Minnesota. He is the author of *The Revolt of the Engineers: Social Responsibility and the American Engineering Profession* (1971).

W. David Lewis is Hudson Professor of History and Engineering at Auburn University. He is the author of several books on aviation history, including (with W. P. Newton) *Delta: The History of an Airline* (1979).

Hugo A. Meier teaches history at Pennsylvania State University. His writings center on concepts of technology in nineteenth-century American social history.

Carroll Pursell is Adeline Barry Davee Distinguished Professor of History at Case Western Reserve University, where he directs the graduate Program in the History of Technology and Science.

Bruce Sinclair is the Melvin Kranzberg Professor of History at the Georgia Institute of Technology. His most recent book is *A Centennial History of the American Society of Mechanical Engineers, 1880–1980* (1980).

Merrit Roe Smith is Professor of the History of Technology in MIT's Program in Science, Technology and Society. His book *Harpers Ferry Armory and the New Technology* (1977), won the Frederick Jackson Turner Award from the Organization of American Historians.

Darwin H. Stapleton is Director of the Rockefeller Archives. He has published several books on the transfer of European technology to America as well as on the engineer Benjamin Henry Latrobe.

John William Ward was formerly President of Amherst College and directed the American Council of Learned Societies.

James C. Williams teaches history at DeAnza College and is Executive Director of the California History Center. His current research is on the history of energy production and use in California.

INDEX

Elwell, Cyril F., 195, 278
Emerson, James B., 103
Energy, Department of, 260
Energy sources
atomic, 248–249, 250
early electrical industry, 242
and mechanization of agriculture, 79
nuclear, 252–255, 259–261
planning for, 247
radioactivity, 250
and REA, 246
science of, 249
shift in, 5
Engineering
Latrobe's impact on, 35
need for education in, 68
scientific method in, 94
Engineering societies, 4
Engineers in management, 166–167
England. *See also* Colonial era
and technology transfer, 41, 44
transportation improvements in, 29, 35
Enlightenment, French, 19
Environmental movement, vs. conservation movement, 151
Ericsson, John, 86
Erskine, John, 212
Europe and technology transfer, 20, 21, 41, 44
Evans, Oliver, 2, 29, 54

Fairchild Camera and Instrument Corporation, 288–290
Faraday, Michael, 108, 118
Farmer, E. E., 277
Farmer, Fannie Merritt, 146
Farragut, Adm. David, 85
Federal Telegraph Co., 197
Fermi, Enrico
and control of nuclear fission, 252–255

intellectual pursuits of, 262
"Fermi statistics," 217
Field, Cyrus W., 80
Fifty Years on the Mississippi (Gould), 81
Fire-suppression programs, in modern forestry, 154
Fitch, John, 2
Fizeau, Armand, 130
Flad, Henry, 87
Fleming, George, 23
Flour mill, automatic, 2
Fonck, Capt. René, 212
Foot, Adonijah, 58
Ford, Henry, 6
advertising of, 183
as folk hero, 188, 189
and impact of automobile, 177–178
manufacturing techniques of, 180–181, 183–184
Model T of, 182, 187, 188
relationship with labor, 185
Ford Motor Company, 180, 181, 187
Forest Service, U.S., 152
Forestry
and conservation movement, 151
environmental, 161
and outdoor recreation, 158–160
scientific, 154, 158, 160
sustained-yield, 153
Foucault, Jean, 130
Fourneyron, Benoit, 95–96
Francis, James Bicheno, 93
as chief engineer, 93–94
experimental testing by, 103–104
hydraulic science of, 98–99
Lowell Hydraulic Experiments, 103
turbine design by, 95–97, 100, 101
Francis, John, 93

Howe, Henry, 48
Hubbard, Gardiner Greene, 110, 116
Humphreys, A. A., 90
Humphreys, Joshua, 15
Hungarian (roller) process, 2
Hussey, Obed, 74
Hydraulic engineering, in California, 276–277
Hydraulic science, 90, 97–98
Hydraulic turbine, 93
Hydro projects, 243, 246

Ickes, Harold L., 242, 245
Image of America (Bruckberger), 177
Incandescent electric lamp, 5
Indians, American
agriculture of, 72
Jefferson on, 21
Industrial Revolution, 1, 2, 10
in agriculture, 71
in America, 13
science fiction as product of, 268
Industrialization
assembly line, 6, 183–184, 223
impact on women's lives, 145
and mechanization of agriculture, 79
and scientific technology, 93
Industry *See also specific types of industry*
application of science to, 139
and R&D, 271
Innovation
continuous, 140–142
encouragement toward, 14
industrialization of, 5
Interchangeable parts
development of, 48
in weapons industry, 50, 56
Internationalism of science, 21

Invention
encouragement toward, 14
Jefferson on, 28–29
Inventors, as folk heroes, 68, 117–118, 128, 174, 175
IQ testing, 279
Iron making
effect of Industrial Revolution on, 2
during wooden age, 9–10
Ironclads, 84. *See also* Shipbuilding
Isaacs, Jacob, 32

Jackson, Andrew, 64
Jefferson, Peter, 17
Jefferson, Thomas, 17, 42, 47, 221
compared with Franklin, 22
correspondence of, 18–19
household technology of, 27
interests of, 23–24
on invention, 28–29
as minister to France, 18
nationalism of, 20–21
as patent administrator, 32–33
and patent law, 30–31
Jeffery, Thomas B., 181
Jensen, Peter L.
invention of loudspeaker, 190–194, 200–203
in later years, 207–209
Magnavox Co. and, 204–205
Jerome, Chauncey, 14
Jet propulsion, development of, 6
Jobs, Steven, 290
Johnson, Hiram, 198
Johnson, J. D., 55
Johnson, Robert, 55
Jones, Thomas P., 62
as head of U.S. Patent Office, 63
and *Journal of Franklin Institute*, 62, 65
and technical education, 68–69

Jordon, David Starr, 278
Journal of Franklin Institute, 62, 65

Keaton, Buster, 229
in *The General*, 230–231
in *The Navigator*, 230
Keller, Helen, 107
Kendall, Edward, 48
Kennett, Luther M., 84
Kentucky River Bridge, 87
King, Judson, 239, 240
Kipling, Rudyard, 219
Klystron, 282, 283
Knox, Gen. Henry, 25
Kodak camera, 135
Kreusi, John, 119, 121, 127

Labor, in "Whitney's American system," 50
Laboratories, industrial research, 5. *See also* Research, industrial
Laborde, Albert, 249
Land-grant institutions, origins of, 70
Langley, Samuel, 114
Langmuir, Irving, 5
Lasswitz, Kurd, 268, 269
Lathe, Blanchard, 12
Latrobe, Benjamin Henry, 34–35
canal engineering of, 40–41
and English personnel, 41–42
move to U.S., 35
as Surveyor of Public Buildings of U.S., 36–37
waterworks of, 37–40
Latrobe, Henry, 42
Lawrence, Ernest, 251
Lee, Roswell, 53, 54, 57, 58
Lenthall, John, 41, 43
Lesseps, Ferdinand de, 91
Levassor, Emile Constant, 178
Ley, Willy, 269

Library of Congress, Jefferson's contribution to, 19
Lindbergh, Charles A., 214
achievements of, 225
early life of, 220–221
first trans-Atlantic flight, 211–215
as national symbol, 215–220
relationship with Goddard, 266, 271, 272, 273
Litton, Charles, 282
Loudspeaker
early demonstrations of, 203–205
invention of, 191, 200–203
Lowell, Proprietors of Locks and Canals, 93, 97
Lowell Hydraulic Experiments, 103
Lowrey, Grosvenor P., 123

McCabe, James D., Jr., 80
McCarthy, Francis J., 195
McCormick, Cyrus Hall, 3, 73
business techniques of, 75
early reapers of, 74
McCormick, Robert, 74
McFarland, J. Horace, 158
McGee, W. J., 156
MacFarland, Jacob Corey, 59
Machine, in cinema, 234–235. *See also* Technology
Machine tool industry, pre-Civil War, 59
Madison, James, 28
Magnavox, 201
Magnavox Co., 205, 206–207
Management techniques, in conservation movement, 155. *See also* Scientific management
Manhattan Project, 255, 256, 261
Manufacturing, "American System" of, 2, 14, 45, 57
Marketing, and technology, 72
Marsden, Ernest, 250

Marston, Obed, 67

Mass production, 1. *See also* Industrialization
Ford's technique of, 184
of Kodak camera, 134

Mathematical theory, application of, 94, 96

Mechanics' institute movement, 70

Mechanization
in agriculture, 72
of arts, 227
farm, 77
19th-cenury movement toward, 16

Mees, C. E. Kenneth, 129, 136, 138–140

Menlo Park, 119, 122

Mesabi range, 126

Metal working, in early America, 55

Meteorological research, and rocketry, 267

Method, and technological success, 224. *See also* Research

Middletown, Connecticut, armory at, 55

Midvale Steel Company, 166, 168

Military
institutional support of, 269–274
R&D funding of, 270
and science and technology, 269, 274–275
weapons development of, 275

Mill Rock, Connecticut, armory at, 48, 50, 51–53

Millikan, Robert, 251

Milling machine, "Whitney," 49

Mills, Robert, 52, 43

Mississippi River, and Eads's jetties, 90

Model T, 177, 182, 183, 187, 188. *See also* Automobile industry

Modern Times (1936), 6, 227, 231–236

Monroe, James, 47

Monticello, 26

Morgan, J. P., 123

Morrill Act (1862), 70

Morse, Samuel F. B., 108

Motion pictures. *See also* Photographic industry
Buster Keaton, 229–231
Charlie Chaplin in, 231–236
chase scenes in, 229
invention of, 228
nature of technology in, 236

Motion study, Gilbreth's, 170–171

Muir, John, 158, 161

Munson, William Giles, 52

Muscle Shoals proposal, 242

Nagasaki, 255

NASA-Ames Research Center, 288

National Advisory Committee on Aeronautics, 5

National Bureau of Standards, 5

National forest policy, 152

National Geographic Magazine, 116

Nationalism, 20

National Park movement, 157

National Park Service, 157

The Navigator, 230

Nelson, William, 83

Neutron, discovery of, 251

New Deal, 242, 246

New Machine, 174–175

Newell, Frederick, 156

Newport, Rhode Island, as shipbuilding center, 10

Nichols, William R., 143, 144

Norris, Sen. George W., 240, 242, 245

North, Simeon, 55, 56, 57, 61

Watertown, Massachusetts, armory at, 175
Waterworks technology, 39–40
Watson, Thomas A., 110
Watt, James, 2, 35
WE (Lindbergh), 213, 215
Wells, H. G., 232, 267–268, 269
West Point, U.S. Military Academy at, 4, 24
Western River Improvement and Wrecking Co., 84
Western Union Telegraph Co., 108, 111
Weston, William, 40, 41
Wheat culture, expansion of, 72–73
Wheelwright, 8
Whistler, Maj. George W., 93
White House, Latrobe reconstruction of, 37
Whitman, Walt, 220
Whitney, Eli, 3, 26, 46, 78
  in arms business, 46–48
  cost estimate method of, 51
  legend of, 49
  and mass production, 61
  Mill Rock Armory of, 48, 50
  relation with employees, 53
Whitney, Willis R., 5
"Whitney" milling machine, 49
Whitworth, Joseph, 60
Willard, Joseph, 20
Wilson, Woodrow, 205–206
Windsor, Vermont, armory at, 60
Wireless communication systems, 278
Wistar, Caspar, 32
Wolcott, Oliver, 46, 47
Women
  and domestic efficiency, 173
  impact of sewing machines on, 3
  and introduction of typewriter, 4
Women's Laboratory, at MIT, 143

Wood
  building with, 11–12
  and clockmaking, 14
  and colonial economy, 9
  and shipbuilding industry, 10–11
Wooden age, 8–16
Woodward, Calvin H., 88
Woodworking machinery, development of, 12
Wozniak, Stephen, 290
Wrecking business, 82, 83

319